Gottfried Glöckner

Gradient HPLC of Copolymers and Chromatographic Cross-Fractionation

With 135 Figures

Springer-Verlag
Berlin Heidelberg NewYork London
Paris Tokyo Hong Kong Barcelona

Professor Dr. Gottfried Glöckner

Technische Universität Dresden
Mommsenstr. 13, O-8027 Dresden, FRG

ISBN 3-540-52739-7 Springer-Verlag Berlin Heidelberg New York
ISBN 0-387-52739-7 Springer-Verlag New York Berlin Heidelberg

Library of Congress Cataloging-in-Publication Data
Glöckner, Gottfried, 1925-Gradient HPLC of copolymers and chromatographic cross-fraction/
G. Glöckner. p. cm. Includes bibliographical references (p.) and indexes.
ISBN 3-540-52739-7 (Berlin : alk. paper).—ISBN 0-387-52739-7 (N.Y. : alk. paper)
1. High performance liquid chromatography. 2. Polymers—Analysis. I. Title.
QD79. C454G58 1991 543'.0894—dc20 90-10274 CIP

Printed in Germany

Typesetting: Thomson Press (India) Ltd, New Delhi
Offsetprinting. Mercedes-Druck, Berlin; Bookbinding: Lüderitz & Bauer., Berlin
2152/3020-543210 – Printed on acid-free paper

Preface

"The problems involved in separating complex macromolecules require understanding not only the chromatographic process but also the physicochemical behavior of the solutes."

This sentence from the pen of Phyllis R. Brown[1], University of Rhode Island, can certainly be applied to synthetic copolymers whose structure is very complex indeed. Thus it may be forgiven that a book on copolymer HPLC has been written not by a trained chromatographer but by someone from the polymer side.

The HPLC of synthetic polymers is often understood to mean only a synonym for size exclusion chromatography. The latter method separates polymers according to the size of the macromolecules and enables the molecular weight distribution of a sample to be evaluated. But as early as 1936, Mark and Saito attempted chromatographic fractionation of cellulose acetate on a charcoal-like adsorbent made from blood. HPLC adsorption chromatography was first applied to copolymer analysis by Teramachi et al. in 1979. Since then, another branch of polymer HPLC has arisen which has the capacity of separating copolymers by composition and enables the chemical composition distribution to be evaluated. The technique requires a suitable elution program and is mainly carried out as gradient elution.

Gradient HPLC of polymers is under intense study in several laboratories all over the world. Since copolymers, modified polymers, or polymer blends are of growing importance, efficient methods for separation according to chemical structure are urgently needed, especially because related methods of measuring structural parameters usually yield an average value but not the distribution within a given sample.

With about ten years experience in the field of polymer HPLC it was my ambition to provide a synopsis of reported investigations, to discuss the methodology which led to success in these attempts, and to introduce a technique called chromatographic cross-fractionation, which enables the complex distribution of a copolymer sample by molecular weight and composition to be evaluated. These are the main parts of the present volume.

[1] in: LC–GC International 2 (1989) No. 12, p. 51

I should not miss the chance to express my deepest gratitude to colleagues and good friends not only in my country but also in The Netherlands and in France, in Czechoslovakia and in the Soviet Union, in Japan and in the United States of America. I am much obliged for information on recent publications in the field, for sending reprints or even manuscripts of the latest papers, for notes on unpublished results and for fruitful discussions. This generous assistance was very helpful in the preparation of the manuscript. I hope it will be for the readers' benefit as well.

Unfortunately, there are some pitfalls in adsorption chromatography of polymers, which may be rather annoying to beginners. In order to make it easier to avoid these difficulties, the main part of the book is introduced by a series of chapters discussing the peculiarities of polymer HPLC. This should help the reader to understand the physicochemical background of the phenomena. A special chapter is added, which is full of experimental hints. So I do hope the book is of interest and of use as well.

Chapter 12 is a glossary of some common terms in chromatography or polymer chemistry. In the text, *italics* indicate that the respective term is explained in the glossary or that its definition is (in a few exceptions) given in the context of the introductory Chaps. 1–3. Definitions show the respective term in bold face.

Dresden, October 3, 1990 Gottfried Glöckner

Table of Contents

Symbols and Abbreviations

a	exponent in KMH equation
A	peak area
A	starting eluent in gradient elution
Ac	acetone
AC	adsorption chromatography
AcN	acetonitrile
AN	acrylonitrile
AU	absorbance unit
B	gradient component of higher elution strength
Bd	butadiene
BA	*n*-butyl acrylate
BMA	*n*-butyl methacrylate
BuCl	*n*-butyl chloride
Bzn	benzene
c	concentration (g/l)
$C_1, C_2, \ldots$	local constants in a formula
CCD	chemical composition distribution
CCF	chromatographic cross-fractionation
CHx	cyclohexane
d_0	pore diameter
d_P	particle diameter
DCE	dichloro ethane
DMA	decyl methacrylate
DMF	dimethylformamide
DMSO	dimethyl sulfoxide
Dx	dioxane
$\Delta E(230)$	absorptivity at wavelength indicated
ELS	evaporative light-scattering (detector)
EMA	ethyl methacrylate
EP	epoxy resin

EPDM	copoly(ethylene/propylene/diene monomer), ethylenepropylene rubber
EtAc	ethyl acetate
EtE	diethyl ether
EtOH	ethanol
F	flow rate (ml/min)
g	molar ratio of constituents in a bipolymer, $g = n_A/n_B$
G	molar ratio in a binary monomer mixture, $G = [A]/[B]$
h	height of a theoretical plate
Hex	*n*-hexane
Hp	*n*-heptane
HPLC	high-performance liquid chromatography
HPPLC	high-performance precipitation liquid chromatography
*i*Oct	*iso*-octane, 2,2,4-trimethylpentane
IR	infrared
k'	capacity factor of a sample, $k' = (t_e - t')/t'$
K	distribution constant
	K_{ads}—in adsorption
	K_{SEC}—in size exclusion
	K_{el}—in electrostatic interaction (repulsion)
K	pre-exponential KMH-constant
k	Boltzmann constant
KMH	Kuhn–Mark–Houwink (equation, constants etc.)
L	length of a column (mm)
LLDPE	linear low-density polyethylene
m_0	sample mass, amount injected (μg)
M	molecular weight
	M_{app}—of an appearent repeat unit
	M_n—number average
	M_0—of a repeat unit
	M_w—weight average
	M_i—of component i (i: A, B, ...)
mAU	milli (absorbance unit)
MA	methyl acrylate
MeAc	methyl acetate
MEK	methyl ethyl ketone, 2-butanone
MEMA	2-methoxyethyl methacrylate
MeOH	methanol
MMA	methyl methacrylate

MW	molecular weight
MWD	molecular weight distribution
n_A, n_B,...	number of moles of components A, B,...
N	plate number
n	carbon number
N_L	Avogadro's number
NP	normal phase
NS	nonsolvent
p	probability
	p_{ij}—in copolymerization of monomer j to a terminating unit i
P	degree of polymerization
PMAN	poly(methacrylonitrile)
PMMA	poly(methyl methacrylate)
PrOH	*n*-propanol
PS	polystyrene
PTFE	polytetrafluorethylene
q	molecular-weight ratio of the constituting units in a bipolymer or in a binary monomer mixture, $q = M_A / M_B$
r	monomer reactivity ratio, $r_A = k_{AA}/k_{AB}$, $r_B = k_{BB}/k_{BA}$
r_s	Stokes' radius
$\langle R^2 \rangle^{0.5}$	end-to-end distance (mean value)
R	gas constant
RI	refractive index
RP	reversed phase
s	standard deviation
S	solute acceleration factor, slope factor in gradient elution
$\langle S^2 \rangle^{0.5}$	radius of gyration (mean value)
S	styrene
S/AN	copoly(styrene/acrylonitrile)
S/BMA	copoly(styrene/*n*-butyl methacrylate)
S/EMA	copoly(styrene/ethyl methacrylate)
S/MA	copoly(styrene/methyl acrylate)
S/MEMA	copoly(styrene/2-methoxyethyl methacrylate)
S/MMA	copoly(styrene/methyl methacrylate)
S/TBMA	copoly(styrene/*tert*-butyl methacrylate)
SEC	size exclusion chromatography
t'	elution time of a non-retained sample, column dead-time
t_e	elution time of a retained sample
t''	net retention time, $t'' = t_e - t'$
t_R	retention time ($= t_e$)
TBMA	*tert*-butyl methacrylate
TCM	trichloro methane, chloroform

Tetra	carbon tetrachloride, tetrachloro methane
THF	tetrahydrofuran
Tol	toluene
TREF	temperature-rising elution fractionation
TT	turbidimetric titration
u	linear velocity
UV	ultra violet
v	copolymer composition (volume fraction)
V	Volume V_e elution volume V_i—interstitial volume V_{mob}—mobile-phase volume V_0—injected volume, sample volume (μl) V_P—pore volume V_R—retention volume ($=V_e$) V_{stat}—volume of stationary phase
VAC	vinyl acetate
VAL	vinyl alcohol
w	measure of copolymer composition (weight fraction of the monomer unit of interest)
W	composition of monomer batch, weight fraction
W	Water
x	copolymer composition (mole fraction), $x_A = n_A/(n_A + n_B)$
X	composition of monomer batch, mole fraction, $X_A = [A]/([A] + [B])$
ε	molar absorptivity
ε_p	porosity
Φ	volume fraction Φ_B—of modifier in gradient elution Φ_{NS}—of nonsolvent at cloud point Φ'_P, Φ''_P—of polymer in gel and sol phase, respectively $\Phi'_{P,x}$, $\Phi''_{P,x}$—of copolymer in gel and sol phase, respectively (P: degree of polymerization, x: composition)
ρ	density
σ_1	surface tension af a liquid
σ	steric hindrance parameter
η	viscosity
χ	Flory–Huggins interaction parameter
$[\eta]$	intrinsic viscosity

1 Introduction

Mother nature alone has the capacity of producing uniform macromolecules. Proteins, nucleic acids, and some special carbohydrates are substances whose molecules are created in structure-controlled biochemical operations. The molecular uniformity and consistency of these materials is prerequisite to ordered structures and life. Aside from these key substances, molecular heterogeneity can be found even in natural polymers, e.g. in cellulose and natural rubber.

Man-made polymers are, for the time being, produced by chain polymerisation (i) or step grow polymerization (ii). Synthetic polymers are heterogeneous in *molecular weight* (MW) due to differences in either (i) the life-time of activated species or (ii) the size of the oligomers which are coupled in each reaction step. Thus, molecular-weight distribution (MWD) is inherent in synthetic polymers.

Copolymers are polymers which are produced from more than one species of monomer [1]. In general, the different units are differently incorporated in the polymer molecules which, in addition to MWD, causes distribution in chemical composition (CC). Distributions in MW and CC are to be expected also in polymers derived from homopolymers by incomplete chemical modifications, e.g. in partially hydrolyzed poly(vinyl acetate) containing both acetate and alcohol groups, or in incompletely esterificated cellulose. Polymers of this kind, although not copolymers by definition [1], are also dealt with ("**quasi-copolymers**") because of far-reaching similarities in properties and problems. In fact, it was cellulose acetate whose fractionation caused Rosenthal and White to coin the term "cross fractionation" in 1952 [2].

1.1 About the Aims and Methods of Cross Fractionation

The complex MW/CCD is a distribution in (at least) two directions. Hence, its experimental evaluation requires separation in more than one direction. The classical approach is based upon the dependence of copolymer solubility on composition and chain length. This two-fold dependence had become apparent in a previous observation reported by Rosenthal and White [2]: the fractionation of cellulose acetate in acetone solution with either *n*-heptane or water as a precipitant yielded fractions which varied both in acetyl content and *intrinsic viscosity* (used as a measure of molecular weight, M). Figure 1.1 gives a schematic picture of what is going on.

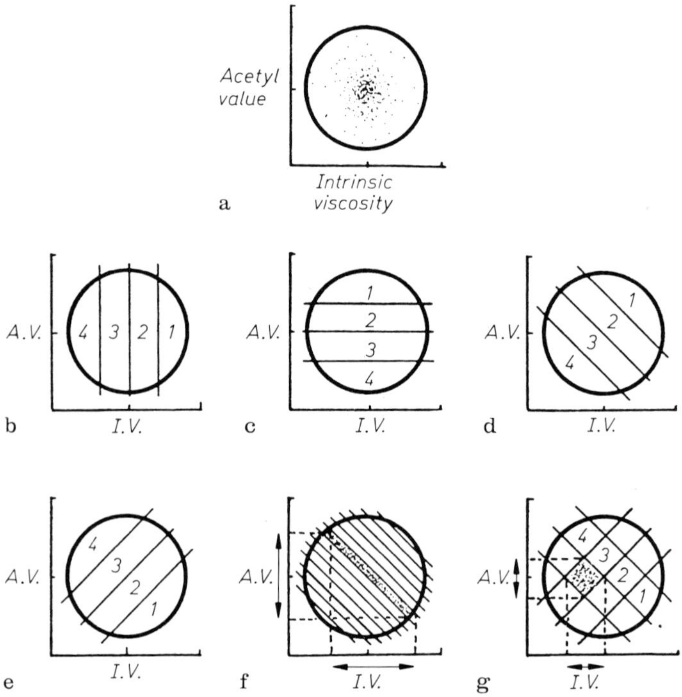

Fig. 1.1 a–g. Fractionation and cross fractionation schemes. (**a**) Two-dimensional distribution of a cellulose acetate sample; ordinate: chemical composition (acetyl content), abszissa: intrinsic viscosity (as a measure of molecular weight); (**b**) fractionation according to MW; (**c**) fractionation according to CC; (**d**) and (**e**): general cases: fractionation simultaneously according to MW and CC, directions determined by the solvent/nonsolvent combination used; (**f**) high-precission fractionation through the solvent/nonsolvent combination used in Fig. 1.1d; (**g**) scheme of cross fractionation with prefractionation by the solvent/nonsolvent system of Fig 1.1d and subsequent fractionation by a complimentary system. From Ref. [2] with permission

A solvent/nonsolvent combination fractionating solely by MW (Fig. 1.1b) would be appropriate for the evaluation of the MWD, another one separating by CC (Fig. 1.1c) would be suited for measuring the CCD of the copolymer. Both MWD and CCD can be understood as cross sections (or profiles) of the distribution figure, parallel to either the composition or the MW axis, respectively. Unfortunately, solvent/nonsolvent systems with separation characteristics of that kind are scarce.

In general, fractionation is influenced by CC as well as by MW (Figs. 1.1d and e). The direction of separation is determined by the experimental conditions, above all by the solvent/nonsolvent combination chosen. A different combination may enable separation in a differing direction. With a given solvent/nonsolvent combination, fine fractionation reduces the amount of material in the fractions but not the heterogeneity of the latter, see Fig. 1.1 f. "It is apparent that in spite of the most painstaking technique each of the fractions would have relatively wide spreads of both intrinsic viscosity and acetyl values" (Rosenthal and White [2]).

Rather homogeneous fractions can be obtained by **cross fractionation** which requires two solvent/nonsolvent systems separating in two different directions (at best, almost at right angles to each other). Fractions of the starting material obtained by one precipitant are redissolved and fractionated once more with the help of the complimentary precipitant yielding the final fractions shown in Fig. 1.1g. (Note that the directions of separation need not run parallel to the axes of the distribution). Reviews on separation and molecular characterisation of copolymers in general have been given by Riess et al. [3], by Inagaki et al. [4] and, on characterization of graft copolymers, by Ikada [5].

Cross fractionation is still as important as it was in 1952. Two-dimensional distributions require two-dimensional separations. The disadvantage of classical cross-fractionation is the amount of labour involved. Separating a sample into 20 to 30 final fractions and measuring all the necessary data requires about three months of skillful work.

It is the purpose of the present title to inform about the advantages of modern techniques for analytical cross fractionation and separation by composition. Although the techniques can be possibly also employed on a preparative scale,

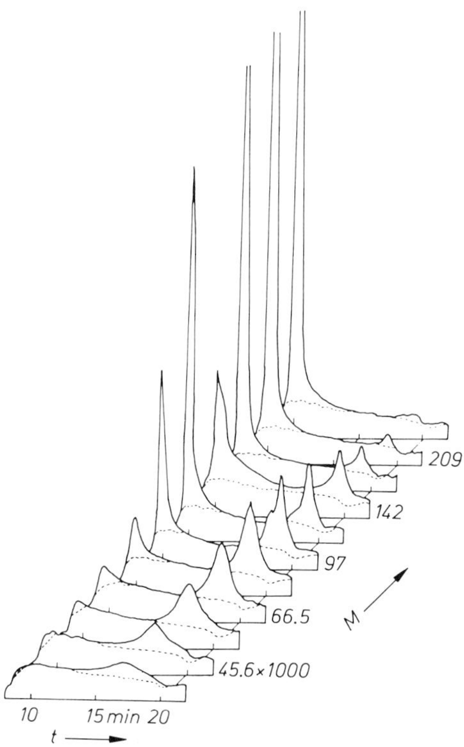

Fig. 1.2. Chromatographic cross-fractionation by gradient HPLC of SEC fractions. Model mixture of two S/AN copolymers, SEC of 0.87 mg starting material on a bank of five columns (300 × 7.8 mm each, packed with μStyragel, $d_0 = 50$ nm, 100 nm, 1000 nm, 10^4nm, and 10^5nm), eluent: tetrahydrofuran, flow rate: 1 ml/min. Each 0.5ml of SEC eluate were collected as a fraction from which aliquotes of 100 – 175 μl were injected into a HPLC apparatus for gradient HPLC on a C18 column (150 × 4.6 mm, LiChrosorb, $d_p = 10\,\mu$m). Gradient *iso*-octane/(THF + 10% methanol); starting eluent: 10% **B**, elution by 60% (at $t = 3$ min) – 90% **B** ($t = 15$ min), multilinear. The first peaks are from the copolymer containing 16.1 mass% acrylonitrile ($M_n = 325{,}000$), the second ones from the copolymer with 30% AN ($M_n = 71{,}000$). By courtesy of Hüthig & Wepf Verlag [9]

the state of the art with respect to preparative separations is, for the time being, still classical fractionation by solubility differences or by distribution in demixing liquids [6, 7]. The latter method is regarded as being advantageous especially in the separation of block and graft copolymers because it diminishes the tendency of micell formation [8].

Chromatographic cross-fractionation (CCF) as an analytic technique applies high performance liquid chromatography (HPLC) to the task in order to make it less time-consuming. Figure 1.2 shows the result obtained on the mixture of two copoly(styrene/acrylonitrile) samples in 1983 [9]. About 1 mg of the mixture was separated by size exclusion chromatography (SEC) into 11 fractions which were subsequently analysed by gradient HPLC. The sample with 16% acrylonitrile (AN) was eluted after about 10 min, the other one eluting between 15 and 20 min was that with 30% AN. As expected from its molecular weight, it was mainly present in the low-MW SEC fractions. The whole separation was completed within six hours.

Thus, much less time was needed for separating a model mixture of copolymers than in analogous experiments performed some years before. Figure 1.3 presents results obtained on a mixture of two S/AN copolymers by column fractionation in 1971 [10]. About two weeks' work was necessary for measuring the data. Comparing Figs. 1.2 and 1.3, note that only two separations by CC are shown in Fig. 1.3 whereas 11 analyses of this kind and an additional separation by molecular size are represented in Fig. 1.2.

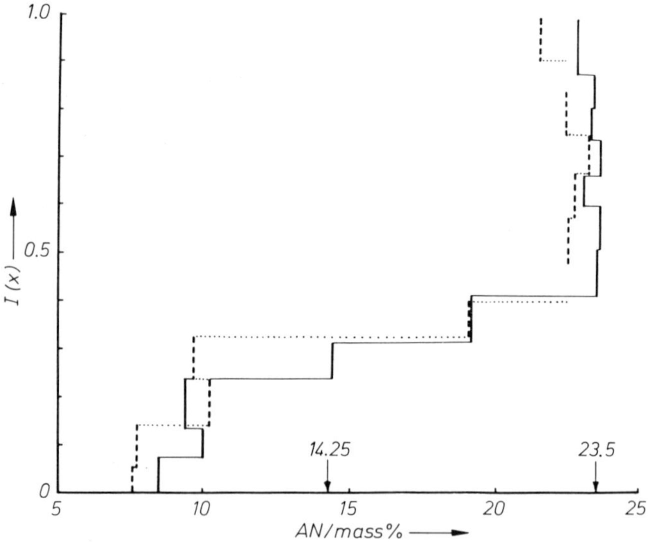

Fig. 1.3. Separation by composition of a 1:1 mixture of two S/AN copolymers by column extraction at 2 °C (*full bars*) or 25 °C (*dashed lines*). Column 1000 × 24 mm, packed with glass beads ($d_p = 300 - 500\ \mu m$), sample load 300 mg, gradient *n*-hexane/dichloromethane. The arrows indicate the composition of the copolymers mixed (weight average). The M_n value of the copolymer with 14.25 mass% AN was 62,000 and that of 23.5% was 44,700 [10]

From Fig. 1.1 it can be concluded that any fractionating procedures, which separate in different directions, can be combined into a cross-fractionating system. It is not necessary that one separation goes strictly by molecular weight and the other strictly by composition, but such an idealized picture allows a straightforward discussion of chromatographic techniques for cross fractionation. We shall take this advantage without forgetting that most separations by composition are to some degree influenced by MW and vice versa. Thus, the guidelines for CCF can be stated by saying that two chromatographic techniques are required, one separating by molecular weight, the other by composition.

Size exclusion chromatography is well known as a method for evaluating MMDs. The process is entropy governed, ideally without any enthalpic interactions between solute molecules and the surface of the packings. Even then, the separation is not strictly due to molecular weight but to hydrodynamic volume, $M\,[\eta]$, where $[\eta]$ represents the *intrinsic viscosity*. Within the limits of this qualification, SEC can be chosen for separation by molecular size.

For CCF, a complimentary chromatographic method is needed which separates by composition. The term "gradient HPLC" represents suitable techniques performed in columns, where retention is due to enthalpic interactions between the solute and the stationary phase. Any unretained (non-excluded) solute is eluted in a volume that corresponds to the total of the interstitial volume and the pore volume of the packing. At the same elution time, the sample solvent emerges from the column, see Fig. 1.4. Interactive retention of a solute causes higher elution times, whereas an excluded solute without interactive retention is eluted with the interstitial volume of the column (see also Fig. 3.3 and the related discussion in Sect. 3.2).

A solvent peak is visible only when the solvent of the sample solution differs from the starting eluent in properties which are recognized by the detector. In *gradient elution*, a sample is usually injected as a solution in a portion of the starting eluent. Then the chromatogram should not show the elution of the solvent plug.

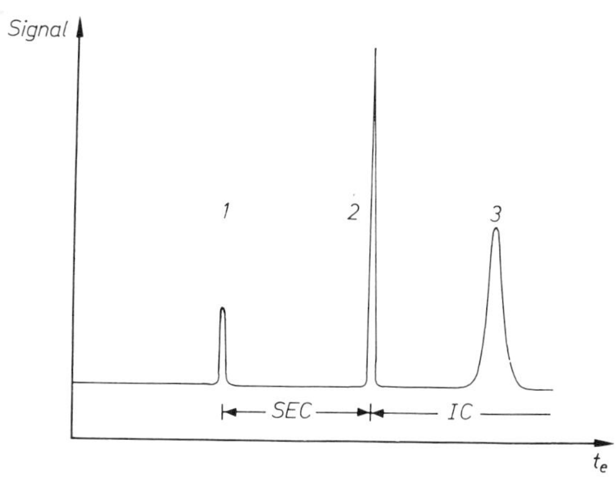

Fig. 1.4. Elution scheme of an incompletely retained polymer sample. *1:* unretained (excluded) portion of the sample polymer, *2:* solvent peak, *3:* properly retained portion of the sample, "SEC": range of SEC elution, "IC": range of elution after interactive retention (unlimited, in principle)

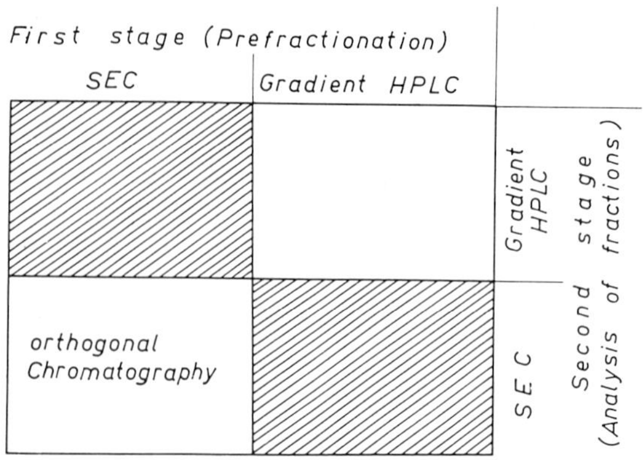

Fig. 1.5. Schematic design of cross fractionation by combinations of HPLC techniques. (*Hatching* indicates promising combinations.)

Both SEC or gradient HPLC can be used in the first or in the second stage of CCF, see Fig. 1.5. The characteristics of polymer HPLC are dealt with in Chaps. 4–6. The separation of copolymers according to composition through gradient HPLC is discussed in Chap. 9. Chapter 10 deals with chromatographic cross fractionation by SEC and gradient HPLC of the fractions.

Most of the experiments dealt with in Chaps. 4–10 were performed with samples dissolved in stabilized THF, whereas the composition of the starting eluent was chosen according to the requirements for proper retention. The reason for the deviation from common practice in gradient HPLC was the need for methods separating by composition, which in the end could be applied to SEC fractions, at best by on-line coupling. THF is the most frequently used eluent in SEC. Thus, stabilized THF was preferred as a sample solvent; a proper gradient

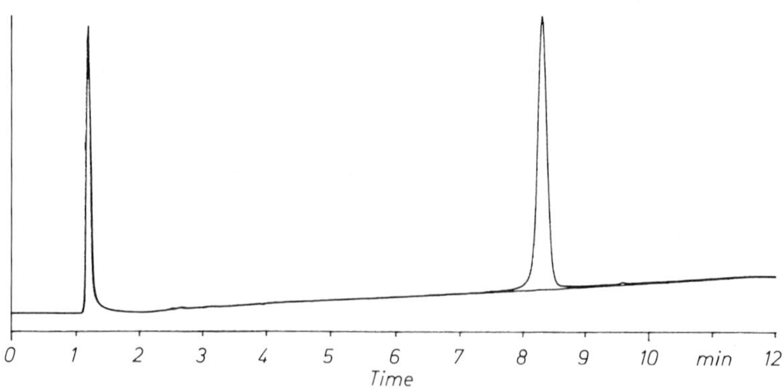

Fig. 1.6. Proper retention and elution of a polymer in gradient HPLC. Column: 60 × 4 mm, CN bonded phase, $d_0 \leq 5$ nm, $d_p = 5$ μm, 50 °C; sample: copoly(α-methylstyrene/acrylonitrile), 15.4 mass% AN, $M_W = 93{,}000$, $m_o = 9$ μg in $V_o = 10$ μl THF; injection into 100% *iso*-octane, gradient: 0–100% THF in 10 min, flow rate: 0.5 ml/min, UV detection at 259 nm, 150 mAU full scale, chromatogram plotted together with an experimental baseline of a blank gradient after injection of 10 μl THF. Solvent peak at 1.2 min, polymer peak at 8.3 min.

elution performed this way will show both the solvent and polymer peak, see Fig. 1.6.

In the present Chapter, a survey is given on papers dealing with classical cross-fractionation (Sect. 1.2) as well as on related work using chromatographic techniques (Sects. 1.3 and 1.4). First of all, however, let us return to the question of whether or not the directions of separation must be parallel to the MW or composition axes of the distribution. For classical cross-fractionation, this question can be denied, *vide supra*, because in classical technique the composition of each fraction as well as its amount and molecular weight are determined experimentally.

The answer is more complicated when data of that kind is derived from the measurement of elution volume. SEC of copolymers is, to some degree, influenced by composition, *vide infra*. Gradient HPLC is mainly separation by composition, but a slight influence of MW must be taken into account as well. Both side effects can be compensated by appropriate calibration. This allows a reasonable approximation to the desired result. The smaller the composition effect in SEC and the MW effect in gradient HPLC are the better is the approximation, see Sect. 3.4.4. The rigorous evaluation of MW and composition requires additional information. A promising approach is prefractionation by gradient HPLC and subsequent analysis of the fractions by multi-detection SEC.

1.2 Classical Cross-Fractionation

In **fractionation by differences in solubility**, a *bipolymer* of *degree of polymerisation*, P, and chemical composition v is subdivided between a polymer-rich gel phase and a more diluted sol phase according to Topchiev et al. [11]

$$\frac{\Phi''_{P,v}}{\Phi'_{P,v}} = \exp[P(\sigma + K\,v)] \tag{1.1}$$

where v indicates the volume fraction of one constituting unit of the copolymer, $\Phi''_{P,v}$ and $\Phi'_{P,v}$ the concentration (volume fraction) of the copolymer in the gel or sol phase, respectively, and σ and K fractionation parameters. Note that $K = 0$ enables a copolymer to be fractionated solely according to MW but theoretically there is no chance of fractionating according to composition without inherent MW effect.

Cross fractionation according to the classical suggestion of Rosenthal and White can be understood via Eq. (1.1) as a sequence of separations with K values of opposite signs. Usually, CF is carried out by solvent/nonsolvent systems. For a binary combination of that kind (components "1" and "2") with an A–B bipolymer, Teramachi and Nagasawa [12] derived the formula

$$K = (\Phi'_1 - \Phi''_1)(\chi_{1A} - \chi_{1B}) + (\Phi'_2 - \Phi''_2)(\chi_{2A} - \chi_{2B}) \tag{1.2}$$

where $\Phi'_{1(2)}$ or $\Phi''_{1(2)}$ are the volume fractions of solvent 1 or 2 in the sol or gel phase, respectively. The parameters χ_{iA} and χ_{iB} describe the interactions between solvent components i ("1" or "2") with the constituting units A or B in a copolymer, respectively. Equations (1.1) and (1.2) are based on the lattice model of Flory–Huggins theory [13–15].

In terms of Eq. (1.2), cross fractionation requires a solvent/nonsolvent system where $\chi_{iA} > \chi_{iB}$ and another one with $\chi_{iA} < \chi_{iB}$ so that K is positive in one system and negative in the other. Teramachi and Kato [16] described the search for systems of that kind by using two bipolymers of different composition. These samples were dissolved in several solvents and titrated by using suitable nonsolvents. This way, the cloud point was measured in a variety of solvent/nonsolvent combinations. Cross fractionation was performed using those combinations which yielded the largest differences in volume fraction of nonsolvent, Φ_{NS}, for the probe copolymers.

As already mentioned in the discussion of Eq. (1.1), the MW distribution of a copolymer can be evaluated by an one-direction fractionation on condition that $K = 0$. Unfortunately, the CC distribution cannot be evaluated in a corresponding manner. Even if repeated one-directional fractionations in different solvent/nonsolvent combinations yield identical CCD curves, these results do not reflect the true distribution of the copolymer. This has been proven experimentally by Teramachi and Kato [16].

From the masses of the individual fractions, m_i, and their composition (weight fraction w_i), the **variance of chemical composition** [17] in the whole copolymer can be calculated:

$$s^2 = \Sigma m_i (w_i - \bar{w})^2 / \Sigma m_i \tag{1.3}$$

where $\bar{w}$ is the mean value. Additional information can be obtained by calculating the **symmetry** of the distribution [17]

$$V = U^+ / U^- \tag{1.4}$$

where (with $w_i > \bar{w} > w_j$)

$$U^+ = \Sigma m_i (w_i - \bar{w}) / \Sigma m_i \tag{1.5}$$

and

$$U^- = \Sigma m_j (\bar{w} - w_j) / \Sigma m_j \tag{1.6}$$

Table 1.1 compiles published results of cross fractionations. Obviously, this fundamental procedure has not become a routine method in a period of more than three decades. It is certainly too tedious for plant control or guidance in the development of new materials. In addition to this, suitable solvent/nonsolvent combinations may not be available in any case.

In Table 1.1, the last four entries deal with the cross fractionation of products obtained by chemical modification of homopolymers. For three of these "quasi-copolymers", the variance in chemical composition, Eq. (1.3), was of the same order of magnitude as that of azeotropic S/MMA [18] or S/MEMA [19] copolymers, but VAC/VAL quasi-copolymer prepared by partial alkaline hydrolysis of PVAC had a variance which was almost threefold as large as that of VAL/VAC prepared by partial acetylation of PVAL.

Table 1.1. Cross fractionation of two solvent/nonsolvent system

Sample	**System 1** Solvent/non-solvent	Direction of separation[a]	No. of intermediate fractions	**System 2** Solvent/non-solvent	Direction of separation[a]	No. of final fractions	**Reference** Year/Ref.
VAC/N-VP[b]	Ac/EtE	[η] ↘ VAC ↗	3	iPrOH/Hex	[η] ↘ VAC ↘	12	1967 [38]
S/Bd	Bzn/MEK	S ↗	4	CHx/iOct	S ↘ *M* ↘	20	1970 [16]
S/MMA(az)[c] (15 g)	Ac/AcN	(MMA ↗)	6	BuCl/CHx	*M* ↘	35	1971 [18]
S/AN (15 g)	MEK/CHx	AN ↘	5	EtCb/EtCN[d]	AN ↗	32	1974 [39]
S/MEMA[e]	Bzn/CHx	S ↗	5	MEK/MeOH	S ↘	22	1978 [40]
S/MEMA(az)	Bzn/CHx	S ↗	6	MEK/MeOH		27	1981 [19]
S/MA (5.06g)	TCM/CHx	S ↗	4	HEMeE[f]/MeOH	S ↘	19	1981 [41]
PET/PTME[k]	Tetra/MeOH	PET ↘	6	TCE[l]/Hp	PET ↘	18	1989 [45]
MMA/MVK[g]	DMF/W + 0.5% NH_4Cl	[η] ↘	3	TCM/EtE	MMA ↗	15	1977 [42]
VAL/VEtCb[h]	THF/Hex	VAL ↘	4	MeOH/ W + 1% NaCl	VAL ↗	18	1983 [43]
VAL/VAC[i]	THF/Hex	*M* ↘	5	MeOH/ W + 0.5% NaCl	*M* ↘, VAC ↗	18	1986 [44]
VAC/VAL[j]	DMF/ Tol + Hex (1:2)	*M* ↘	4	Ac + W(3:2)/ W + 0.5% NaCl	*M* ↘, VAC ↘	16	1986 [44]

Lokal abbreviations:
[a] on reduction of dissolution power;
[b] N-VP: N-vinyl pyrrolidone;
[c] (az): azeotropic copolymer;
[d] EtCb: ethylene carbonate; EtCN: ethylene cyanohydrine;
[e] model mixture, 10 characterized copolymers;
[f] HEMeE: 2-hydroxyethyl methyl ether;
[g] MVK: units of PMMA, β-functionalized by $(CH_3)_2NSO_2CH_2Li$;
[h] VEtCb: vinyl ethyl carbonate, by modification of PVAL with ethyl chloroformate;
[i] by partial acetylation of PVAL;
[j] by partial alkaline hydrolysis of PVAC;
[k] poly(ethylene terephthalate)/poly(tetramethylene ether) multiblock copolymer;
[l] trichloro ethylene.

1.3 Cross Fractionation Including Size Exclusion Chromatography

The time spent on the evaluation of distribution in MW and CC can be reduced by, e.g. combining any separation by composition with SEC of the fractions. If the first process is operative without substantial dependence on MW, this combination of methods can be considered an approximation towards the most straightforward scheme of cross fractionation, i.e. to a combination of separations parallel to the respective axes of the MW/CCD plot.

1.3.1 SEC Analysis of Fractions Obtained by Precipitation or Extraction

Although Eq. (1.1) predicts that solubility-based fractionation cannot separate only by composition, the MW effect may be small in favourable cases. Indeed, successful analyses of the complex distribution-surface of copolymers have been performed by fractionating precipitation and SEC analysis of the fractions [20, 21] or by column elution and SEC [22].

Fractionation by solubility differences in combination with SEC of the fractions is also the basis of an automated apparatus for the two-dimensional analysis of olefinic copolymers which was introduced by the authors as "cross-fractionating chromatograph" [23], see Sect. 10.5.

Extraction by hydrocarbon supercritical fluids has been used for fractionating low-MW copolymers. Subsequent analyses by dual-detection SEC of the fractions obtained from bipolymers of styrenic and acrylic monomers showed that the styrene content increased with increasing MW [24].

1.3.2 Preparative Thin-Layer Chromatography and SEC Analysis of the Fractions

Thin-layer chromatography (TLC) of polymers has been reviewed by Inagaki [25, 26] and Belenkii and Gankina [27]. The combination of TLC and SEC according to the principles of cross fractionation has been reported repeatedly.

Poly(methyl methacrylate-*b*-styrene-*b*-methyl methacrylate) samples were investigated by TLC with subsequent analysis of isolated portions by SEC [28].

Ascending adsorption chromatography in a dry column packed with silica and SEC analysis of fractions extracted from the packings of the dismantled column was used in the investigation of a diblock copolymer of the same monomer system [29, 30]. Although not a planar chromatographic method, the prefractionation resembled TLC because the unwetted column soaked up the developing mixture by capillary forces. The two-dimensional fractionation revealed the pronounced chemical heterogeneity of the sample [30], see Fig. 10.2.

1.3.3 Preparative SEC and TLC Analysis of the Fractions

A *stat*-copoly(styrene/methyl acrylate) sample was repeatedly fractionated by SEC. Corresponding fractions were collected and analysed on TLC plates according to the distribution of methyl acrylate units. This procedure is complimentary to that described in the proceeding section and enabled the contour-line map of the sample under investigation to be evaluated [31]. The method combination has been applied also to statistical copolymers of S/EMA [32, 33].

1.4 Cross Fractionation by Coupled SEC Techniques

Since *intrinsic viscosity* is influenced by the solvent and SEC separates according to hydrodynamic volume, SEC in different eluents can separate a copolymer in two diverging directions. This is the principle of "**orthogonal chromatography**" suggested by Balke and Patel [34]. The authors coupled two SEC instruments together so that the eluent from the first one (SEC #1) flowed through the injection valve of the second one (SEC #2). At any desired retention time, the flow through SEC #1 could be stopped and an injection made into the second instrument. The first instrument (housing up to 12 PS-gel SEC columns) was operated with THF as an eluent, the second one (with three μ-Bondagel polyether bonded-phase columns) with mixtures of THF and *n*-heptane (Hp).

The authors reported the investigation of statistical copoly(styrene/*n*-butylmethacrylate) samples obtained by free radical polymerisation and the separation of mixtures which contained a copolymer of that kind (S/BMA) together with the parent homopolymers, PS and PBMA. The concentration of Hp in the eluent of SEC #2 had a striking influence on the retention: with pure THF, the mixture of the polymers mentioned eluted in an uniform band at about 700 s which was the SEC elution time. This behaviour remained almost unchanged on the addition of Hp up to 50% although the latter is a nonsolvent for PS. At 60% Hp the PS eluted in a separate peak at about 900 s, see Fig. 1.7. At carefully adjusted compositions between 62.5 and 64.3% Hp, three peaks were

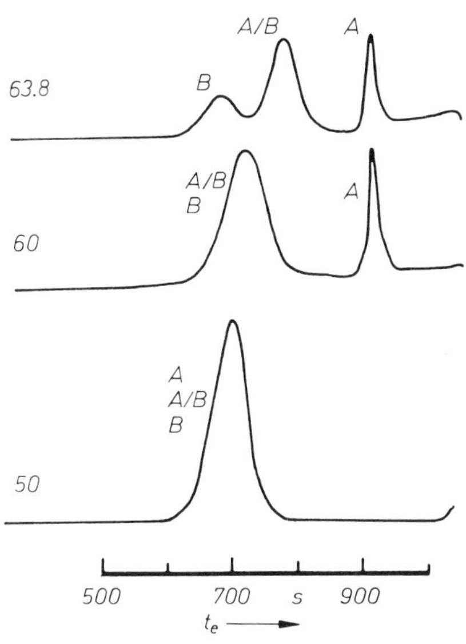

Fig. 1.7. Orthogonal chromatography of the mixture of "*A*" polystyrene, "*B*" poly(*n*-butyl methacrylate), and "*A*/*B*": azeotropic copoly (S/BMA). Isocratic elution in THF/*n*-heptane mixtures (vol% Hp indicated) on a bank of three polyether bonded-phase columns. From Ref. [36] with permission

observed; the elution order was PBMA < S/BMA < PS. Optimum resolution was obtained at 63.8% Hp.

Balke correctly pointed out [35] that the styrene-rich molecules will shrink due to the addition of nonsolvent and thus enter more pores of the column packing than the molecules rich in BMA.

Dawkins repeated these experiments with either a PS gel or a poly(acryl amide) gel column (PAA) in SEC #2 [37]. With the PS gel column and THF:Hp mixtures, the results were almost identical with the observation made by Balke: the mixture of PBMA with PS and a S/BMA copolymer (65.1 mol % S) showed the first signs of separation at 63% Hp; at 64.5% the PS peak was separated from the rest. At 65% Hp, the PS peak almost merged with the solvent peak, whereas at 67% Hp the copolymer was separated from PBMA homopolymer.

Hence, a polyether bonded-phase column [34] and a PS gel column [37] enable corresponding separations to be performed in almost the same range of mobile phase composition (60–64.3% or 63–67% Hp). Things were different in the PAA gel column where polymers were obviously adsorbed when the eluent contained more than 40% Hp. Basically the same difference between the PS and PAA column was observed with THF:*iso*-propanol mixtures [37].

Another example of orthogonal chromatography has been reported recently: the mixture of poly(lauryl methacrylate), PS, and poly(ethyl methacrylate) was separated by SEC #2 using THF:Hp mixtures (55–58% Hp) with the components eluting as individual peaks in the given order [35].

Note that all examples of orthogonal chromatography show elution of the polymer in the exclusion range, i.e. before the solvent peak. This might be the reason why separation is achieved without an external elution gradient. (Balke points out that the plug of sample solvent THF may cause an effective gradient as it sweeps through the columns of SEC #2 [35].)

The separation in conventional SEC #1 is separation by hydrodynamic volume, i.e. it is effected by MW and composition. It is unlikely that the separation in SEC #2 occurs really in an orthogonal direction. But even without literal meaning the term coined by Balke and Patel is an useful denotation of cross fractionation by means of coupled SEC apparatus. The resolution power of this CCF technique is limited by the rather small degree of modification in hydrodynamic volume which can be achieved by changing the solvent. In this respect, interactive chromatography with gradient elution is obviously superior, see, e.g. Sect. 9.7.

1.5 Elution Under Critical Conditions

The elution behaviour of polymers on columns with porous, active packings can change from SEC elution to retention by adsorption on changing the composition of a mixed eluent. The transition is rather sensitive to temperature and occurs in a narrow range of composition which is called the critical composition. Here, retention is independent of molecular weight and *isocratic elution* yields separation according to composition, see Sect. 6.1, Figs. 6.2 and 6.3.

1.6 References

1. Jenkins AD, Loening KL (1989) In: Booth C, Price C (eds) Comprehensive polymer science, vol 1, Pergamon, Oxford, p 34
2. Rosenthal AJ, White BB (1952) Ind Engin Chem 44: 2693
3. Riess G, Callot P (1977) In: Tung LH (ed) Fractionation of synthetic polymers, Marcel Dekker, New York, p 445
4. Inagaki H, Tanaka T (1982) In: Dawkins JV (ed) Developments in polymer characterization, vol 3, Applied Science Publishers, London, p 1
5. Ikada Y (1978) Advances in Polym Sci 29: 47
6. Kuhn R (1976) Makromol Chem 177: 1525
7. Kuhn R (1980) Makromol Chem 181: 725
8. Kratochvil P, Stejskal J, Procházka O, Tuzar Z, Podesva J (1987) 31st IUPAC Macromolecular Symp, Merseburg, June 30–July 4, Lecture IV/IL/4
9. Glöckner G, Koningsveld R (1983) Makromol Chem Rapid Commun 4: 529
10. Glöckner G, Francuskiewicz F, Müller KD (1971) Plaste und Kautschuk 18: 654
11. Topchiev AV, Litmanovich AD, Shtern VYa (1962) Dokl Akad Nauk SSR 147: 1389
12. Teramachi S, Nagasawa S (1968) J Macromol Sci Chem A2: 1169
13. Flory PJ (1942) J Chem Phys 10: 51
14. Huggins ML (1942) J Phys Chem 46: 151
15. Flory PJ (1953) Principles of polymer chemistry, Cornell University Press, Ithaca, New York, p 495
16. Teramachi S, Kato Y (1970) J Macromol Sci Chem A4: 1785
17. Cantow HJ, Fuchs O (1965) Makromol Chem 83: 244
18. Teramachi S, Kato Y (1971) Macromolecules 4: 54
19. Stejskal J, Kratochvil P, Straková D (1981) Macromolecules 14: 150
20. Fritzsche P (1965) Faserforsch Textiltechnik 16: 466
21. Fritzsche P, Klug P, Gröbe V (1971) Faserforsch Textiltechnik 22: 250
22. Ogawa T, Sakai M (1981) J Polym Sci A-2, Polym Phys 19: 1377
23. Gotoh Y, Usami T, Takayama S (1988) Poster presentation, 1st ISPAC Meeting, Toronto, June 2–3
24. Scholsky KM, O'Connor KM, Weiss CS, Krukonis VJ (1987) J Appl Polym Sci 33: 2925
25. Inagaki H (1977) Advances in Polym Sci 24: 189
26. Inagaki H (1977) In: Tung LH (ed) Fractionation of synthetic polymers, Marcel Dekker, New York, chap 8
27. Belenkii BG, Gankina ES (1977) J Chromatog 141: 13
28. Belenkii BG, Gankina ES, Nefedov PP, Lazareva MA, Savitskaya TS, Volchikhina MD (1975) J Chromatog 108: 61
29. Tanaka T, Omoto M, Donkai N, Inagaki H (1980) J Macromol Sci Phys B17: 211
30. Inagaki H, Tanaka T (1982) Pure Appl Chem 54: 309
31. Teramachi S, Hasegawa A, Yoshida S (1983) Macromolecules 16: 542
32. Tacx JCJF, German AL (1989) J Polym Sci A-1, Polym Chem 27: 817
33. Tacx JCJF, German AL (1989) Polymer 30: 918
34. Balke ST, Patel RD (1980) J Polym Sci, B, Polym Letters 18: 453
35. Balke ST (1987) ACS Symposium Series 352, J Amer Chem Soc, p 59
36. Balke ST (1982) Sep Purific Methods 11: 1
37. Dawkins JV, Montenegro AMC (1989) Britisch Polym J 21: 31
38. Agasandyan VA, Kudryavtseva LG, Litmanovich AD, Shtern VYa (1967) Vysokomol Soedin A9: 2634
39. Teramachi S, Fukao T (1974) Polym J 6: 532
40. Stejskal J, Kratochvil P (1978) Macromolecules 11: 1097
41. Teramachi S, Hasegawa A, Hasegawa S, Ishibe T (1981) Polym J 13: 319
42. Bourguignon JJ, Bellissent H, Galin JC (1977) Polymer 18: 937
43. Arranz F, Sánches-Chaves M, Molinero A, Martinez R (1983) Makromol Chem, Rapid Commun 4: 297
44. Arranz F, Sánches-Chaves M, Riofrio A (1986) Makromol Chem 187: 1215
45. Xu Z, Yuang P, Zhong J, Jiang E, Wu M, Fetters LJ (1989) J Appl Polym Sci 37: 3195

2 Chemical Heterogeneity of Copolymers

Chemical heterogeneity is a consequence of chemical composition distribution (CCD) which can be presented as, e.g. an integral or differential distribution curve. Information on the complex MW/CCD can be provided as shown in Fig. 2.1a and b by plotting w vs M. In Fig. 2.1a is, for a *bipolymer*, the average molecular weight due to all B units in a macromolecule, M_B, plotted vs the corresponding average M_A of all A units. This mode of plotting [1] is especially suited for the characterization of *block copolymers*, whereas Fig. 2.1b is more advantageous with *statistical copolymers*. In both forms of diagrams, the distribution surface can be indicated by contour lines projected onto the M_B vs M_A or x vs M plane.

In Fig. 2.1a, a straight connection between equal intercepts on the M_A and M_B axes (e.g. line 1) represents copolymers of varying composition but constant MW. As already mentioned, separation by composition without MW dependence cannot be gained by solubility-based fractionation. Thus, a common separation by composition is better represented by line 2 than by line 1. Any straight line through the origin of the diagram, e.g. line 3, is the locus of molecules which are equal in composition, i.e. which have the same A:B ratio. A line of that kind represents separation exclusively by MW. A line perpendicular to one of the axes, e.g. $M_A = \text{const.}$, line 4, is characteristic of separation according to content in B units. In this case, fractions are neither homogeneous in composition nor in MW; they are uniform in number of A units and graded in B content.

In Fig. 2.1b, the lines 1–3 have the same meaning as in Fig. 2.1a. The area *abcd* schematically indicates a MW/CC distribution. The lines *ad* and *bc* connect points of equal MW in both diagrams, whereas *ab* and *cd* connect points of equal composition.

Consider a *statistical copolymer* which has been polymerized in homogeneous reaction from a mixture of A and B monomers. In general, even macromolecules grown under such fortunate circumstances will differ in chemical structure. There are differences in the sequence of the A and B units along the macromolecules, differences in the average chemical composition of the copolymer molecules formed at any instant of the polymerization (instantaneous heterogeneity), and differences due to the depletion of the reaction mixture in one of the monomers. The latter effect, the conversion heterogeneity, will be zero only with *azeotropic copolymers* or with systems where the *monomer reactivity ratios* are equal to 1; the conversion heterogeneity will be small at, e.g. 5% monomer consumption and increase with the *degree of conversion*.

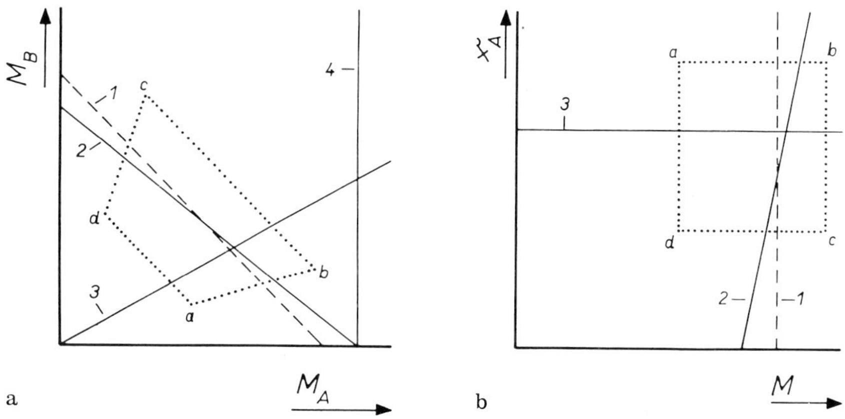

Fig. 2.1 a, b. Graphical representation of copolymer composition and molecular weight (MW) and of distribution in chemical composition (CC) and MW. (**a**) MW due to B units vs MW due to A units, (**b**) mole fraction of A units vs MW of copolymer. In both graphs, line "1" represents molecules of constant MW irrespective of composition, line "2" indicates a scheme of separation according to composition, and line "3" represents bipolymers of constant composition irrespective of MW. The dotted figure *abcd* is the schematic presentation of a given hypothetic MW/CCD in both modes of plotting

Sequence distribution, instantaneous heterogeneity and conversion heterogeneity are dealt with in the following sections under the assumptions that (1) only the ultimate unit of a growing molecule effects the addition of the next monomer, that (2) monomer addition is irreversible and that (3), for all monomeric species involved, the mechanism of addition is the same (e.g. free radical polymerization, ionic polymerization, etc.).

2.1 Sequence Distribution

The *copolymerization propagation probabilities*, p_{AA} and p_{BB}, determine the number fraction distribution $h(z_A)$ and $h(z_B)$ of microblocks formed by z units of type A or B:

$$h(z_A) = (1 - p_{AA}) \cdot p_{AA}^{(z-1)} \tag{2.1}$$

$$h(z_B) = (1 - p_{BB}) \cdot p_{BB}^{(z-1)} \tag{2.2}$$

With the help of the *recursive Mayo–Lewis equation*, the propagation probabilities can be estimated from the chemical composition of a copolymer, and consequently also the *number average* length of uninterrupted sequences of A or B units, L_A or L_B, respectively:

$$L_A = \frac{1}{1 - p_{AA}} = 1 + r_A G \tag{2.3}$$

$$L_B = \frac{1}{1 - p_{BB}} = 1 + r_B/G \tag{2.4}$$

where r_A and r_B are the *monomer reactivity ratios* and $G = [A]/[B]$ the mole ratio in the monomer mixture. For calculation of G from copolymer composition, see Eqs. (12.40 and 12.41).

Figures 2.2a and b present propagation propabilities calculated in this manner for S/AN, S/MMA, and MMA/MA copolymers. Figures 2.3a and b are, for the same copolymer systems, plots of average sequence length vs mole fraction of monomer A in the monomeric batch (2.3a) or in the copolymer (2.3b). Note that, for MMA/MA, S/MMA, and S/AN, the curves of L_A and L_B vs copolymer composition almost coincide. This is due to the general relation

$$L_A = L_B \frac{x_A}{1 - x_A} \tag{2.5}$$

and (apart from systems with very large values of monomer reactivity ratios) to a predominant effect of $x_A/(1 - x_A)$ on the shape of the curves.

The relations between composition of a given copolymer and sequence length parameters demonstrate that separation by composition of a *statistical copolymer* is equivalent to separation by average block length or total sequence distribution. Only with *block* or *random copolymers* and "quasi-copolymers", the sequence length distribution is independent of composition.

The microblock distribution is a distribution along a polymer molecule. Macromolecules of statistical copolymers, even if identical in chain length and composition (and thus also in average sequence length) still offer a great variety with respect to the order of individual microblocks in the molecules. It is certainly not too much an exaggeration if one assumes that a high-molecular-weight sample of a synthetic copolymer will scarcely have macromolecules which are identical in all constitutional parameters. Thus, a copolymer sample contains a tremendous number of constituents. In terms of liquid chromatography, a sample

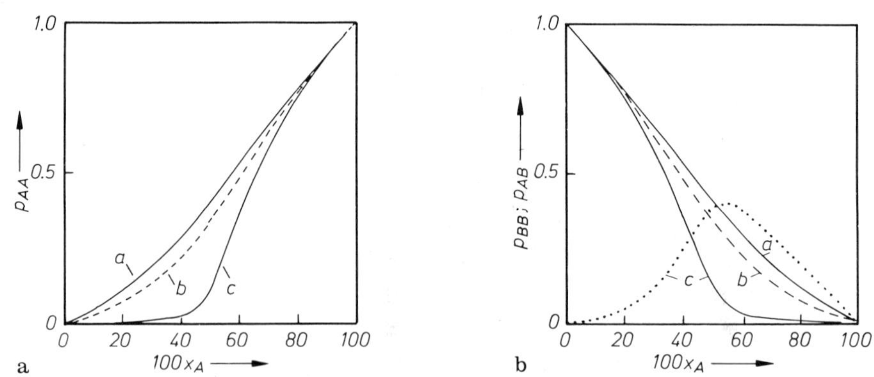

Fig. 2.2 a, b. Effect of copolymer composition on propagation probabilities p_{AA} and p_{BB} for bipolymers "*a*": MMA/MA ($r_A = 0.3$, $r_B = 1.5$; monomer A: methyl methacrylate; "*b*": S/MMA ($r_A = 0.53$, $r_B = 0.49$; monomer A: styrene; "*c*": S/AN ($r_A = 0.41$, $r_B = 0.04$; monomer A: styrene). The probability for the addition of a B unit to a terminating A unit, $p_{AB} = x_B - p_{BB}$, is indicated by the dotted line in Fig. 2.2b for S/AN as an example

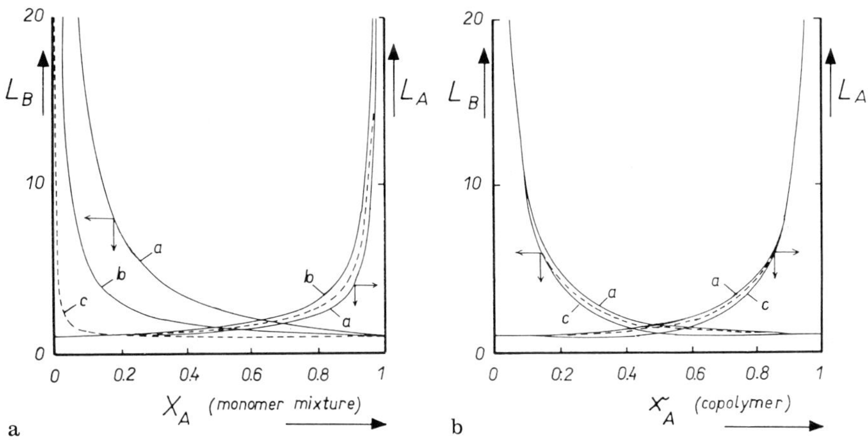

Fig. 2.3 a, b. Average sequence length vs mole fraction of monomer A in the monomeric batch (*left diagram*) or in the copolymer (*right diagram*). Copolymers: "*a*": MMA/MA; "*b*": S/MMA (*dashed curve* in the right diagram); "*c*": S/AN (*dashed curve* in the left diagram)

of this kind is an extremely complex mixture, difficult to separate by size-exclusion or (possibly even more) interactive chromatography.

The average sequence lengths, L_A and L_B, can be measured by physical or chemical methods. The former (e.g. IR or NMR analyses) usually measure the percentage of A or B units inside of triades, pentades, or heptades whereas the latter methods evaluate the percentage of A–A, A–B, and B–B linkages.

The **run number** R is a key in comparing the results of different investigations [2, 3]. By definition, R is the average number of microblocks in a polymer chain of 100 constituting units. It is related to the average sequence lengths L_A and L_B by

$$R = \frac{200}{L_A + L_B} \tag{2.6}$$

and to the *monomer reactivity ratios* and the composition of the monomer batch by

$$R = \frac{200}{2 + r_A[A]/[B] + r_B[B]/[A]} \tag{2.7}$$

Since the sequence length distribution is a distribution within (or along) the macromolecule, vide supra, separation according to (individual) microblock length is possible only after chain scission, e.g. by pyrolysis. Separation of statistical copolymers according to total sequence length distribution is identical with separation by composition.

2.2 Instantaneous Heterogeneity

A general theory of copolymerization has been formulated by Simha and Branson [4]. Since the equations given by them are unfortunately too complex for convenient use, Stockmayer repeated the calculation and published in 1945 a classical paper on the distribution of chain length and composition in copolymers [5]. The calculation has been performed for a binary free-radical copolymerization of monomers A and B. It starts with the balance equations for radicals A* and B* and considers the change in molar concentration with time, $d[A^*_{a,b}]/dt$ or $d[B^*_{a,b}]/dt$:

$$\frac{d[A^*_{a,b}]}{dt} = \underset{\text{(i)}}{k_{AA}[A^*_{(a-1),b}][A]} + \underset{\text{(ii)}}{k_{BA}[B^*_{(a-1),b}][A]} \\ - \underset{\text{(iii)}}{k_{AA}[A^*_{a,b}][A]} - \underset{\text{(iv)}}{k_{AB}[A^*_{a,b}][B]} - \underset{\text{(v)}}{v_{A,t}} \tag{2.8}$$

(The notations $A^*_{a,b}$ and $B^*_{a,b}$ mean that the growing ends of the radicals are formed by an A or B unit, respectively, and that each radical comprises in total "a" units of monomer A and "b" units of monomer B.)

In Eq. (2.8) the contributions (i) and (ii) are the rate expressions for the formation of $A^*_{a,b}$ radicals by addition of A monomer to a corresponding precursor radical. The consumption of $A^*_{a,b}$ radicals is (iii) due to further chain propagation by addition of either A or (iv) B monomer. The last term (v) describes the velocity of the conversion of $A^*_{a,b}$ radicals into stable polymers; it is proportional to the concentration $[A^*_{a,b}]$ of these radicals.

A corresponding rate equation, $d[B^*_{a,b}]/dt$, can be derived for $B^*_{a,b}$ radicals [5]. In the stationary state, both expressions are zero. The solution yields complex sums [4] which have been approximated by integrals and evaluated with the help of Stirling's formula [5]. With the composition (mole fraction) of an individual macromolecule,

$$x_{i,A} = 1 - x_{i,B} = \frac{n_{A,i}}{n_{A,i} + n_{B,i}} \tag{2.09}$$

and the over-all composition $x_A = 1 - x_B$, the composition deviation y of an individual macromolecule from the mean value is defined by:

$$y = x_{i,A} - x_A = x_B - x_{i,B} \tag{2.10}$$

Now, the logarithm of $[A^*_{a,b}]$ or $[B^*_{a,b}]$ can be expanded as a power series in y. This procedure leads to a Gaussian distribution about the mean value which is identical for $A^*_{a,b}$ and $B^*_{a,b}$ radicals. According to Stockmayer, the final result is most compactly expressed as a continuous weight distribution with equal molecular weight assigned to both types of monomer units. If termination is by disproportionation, each radical forms a molecule of same length and composition. The two-dimensional differential weight distribution of degree of polymerization and composition deviation can be written as [5]

$$H(P, y) = \left[\frac{P}{P_n^2}\exp\left(-\frac{P}{P_n}\right)\right]\left[\frac{1}{s(2\pi)^{1/2}}\exp\left(-\frac{y^2}{2s^2}\right)\right] \tag{2.11}$$

where P is the *degree of polymerization* and P_n is its *number average* value.

The parameter s is defined by

$$s^2 = \frac{x(1-x)\kappa}{P} \tag{2.12}$$

where $x = x_A$ and

$$\kappa = [1 - 4x(1-x)(r_A r_B - 1)]^{1/2} \tag{2.13}$$

Equation (2.11) formally consists of two factors. The first one reflects the weight distribution of P for the reaction product of a polymerization with termination by disproportionation. (This kind of distribution is referred to as the *most probable distribution.*) The second factor represents a Gaussian distribution which states that the deviation in composition, y, is normally distributed with standard deviation s.

As Eq. (2.12) shows, the variance s^2 is inversely proportional to P. Thus, the instantaneous composition heterogeneity will be large in low-molecular-weight fractions and diminish in high-MW portions. This effect is to be expected under conditions where the composition of the monomeric mixture does not change in the course of the polymerization, i.e. with the products of low conversion copolymerizations or with *azeotropic copolymers.*

By integration of Eq. (2.11) over all degrees of polymerization, Stockmayer also derived the marginal distribution of compositional deviations [5]

$$\mathrm{H}(y)\,\mathrm{d}y = \frac{3\,\mathrm{d}z}{4(1+z^2)^{5/2}} \tag{2.14}$$

where

$$z^2 = \frac{P_n y^2}{2x(1-x)\kappa} \tag{2.15}$$

2.3 Modification of the Stockmayer Distribution

As already mentioned, Eq. (2.11) was obtained under the condition of (i) termination by disproportionation and (ii) equal MW of both monomer species, A and B. The following passages deal with efforts to overcome these limitations.

2.3.1 Generalization of the Marginal Composition Distribution

For termination by recombination (or mixed termination), the marginal distribution of compositional deviations was derived by Kuchanov [6] and by Stejskal et al. [7]. The result has a form similar to Eq. (2.14) and the graphic representation of both distributions is almost identical. Stejskal and Kratochvīl [8] further generalized Eq. (2.11) by substituting the *Schulz–Zimm distribution* [9, 10] for the *most probable chain-length distribution* and by expressing the mole fraction x_A by the respective weight fraction $\bar{w}_A$ of A units in a copolymer. The ratio of monomer molecular weight is $q = M_A/M_B$.

This way, both limitations (i) and (ii) have been overcome. Integration over all degrees of polymerization yields the differential weight distribution of chemical composition

$$H(w_i - w) = \frac{\Gamma(a + 3/2)}{\Gamma(a + 1)\Gamma(1/2)} \cdot \frac{1}{(1 + z^2)^{a+3/2}} \cdot \frac{dz}{dw_i} \tag{2.16}$$

where

$$z^2 = \frac{P_n q}{2a\kappa w(1 - w)} \cdot \frac{(w_i - w)^2}{[(1 - q)w_i + q]^2} \tag{2.17}$$

$$\kappa = \left\{1 + \frac{4qw(1 - w)\cdot(r_A r_B - 1)}{[(1 - q)w + q]^2}\right\}^{1/2} \tag{2.18}$$

$$\frac{dz}{dw_i} = \left(\frac{P_n q}{2a\kappa w(1 - w)}\right)^{1/2} \cdot \frac{(1 - q)w + q}{[(1 - q)w_i + q]^2} \tag{2.19}$$

Here, $w_i = w_{iA}$ is the weight fraction of A units in an individual macromolecule and $q = M_A/M_B$ is the ratio of monomer molecular weight. $\Gamma(a + 1)$ is the *gamma function* corresponding to $(a + 1)$. Since $\Gamma(1 + 3/2) = (3/2)(1/2)\Gamma(1/2)$, Eq. (2.16) reduces to Eq. (2.14) for $a = 1$ and $q = 1$.

2.3.2 Generalization of Stockmayer's Distribution for Copolymers from Monomers with Different Molecular Weight

Tacx et al. [11] suggested that the Stockmayer function $H(P, y)$ given by Eq. (2.11) should be modified as

$$H'(P, y) = H(P, y) \cdot V(y) \tag{2.20}$$

The function

$$V(y) = 1 + \frac{(q - 1)(x_i - x)}{(q - 1)x + 1} \tag{2.21}$$

can be rewritten as

$$V(x_i) = \frac{x_{iA} M_A + x_{iB} M_B}{x_A M_A + x_B M_B} \tag{2.22}$$

Thus, $V(y)$ is the ratio of the *apparent MW of a repeat unit* in an individual copolymer molecule ($M_{app,i}$, composition x_i) to M_{app} in the whole copolymer of average composition x.

The same correction applies to the modification of the integrated distribution functions (2.14) and (2.16). Stejskal and Kratochvil compared the results of both approaches and found that the difference between $H(w_i - w)$, Eq. (2.16) and $H'(y) = H(y)V(y)$ may become important when the chemical composition distribution is broad, i.e. if the degree of polymerization is low and q differs substantially from unity [8].

2.4 Conversion Heterogeneity

Only under special circumstances is the composition of a copolymer identical with the composition of the monomer batch from which it polymerized. These extraordinary cases are *azeotropic copolymers* or systems whose *monomer reactivity ratios* equal unit, $r_A = r_B = 1$.

In general, the instantaneous composition of a copolymer differs from the composition of the monomer mixture; this causes depletion of the batch in the monomer which is preferably incorporated. Thus, subsequent portions of a copolymer sample are polymerized from mixtures of various compositions; this gives rise to additional chemical heterogeneity. A copolymer with only instantaneous heterogeneity can be obtained only at infinitesimal conversion.

A *copolymerization diagram* (see Fig. 2.4) enables the conversion heterogeneity to be discussed in a graphic manner. Assume the initial composition of the monomer mixture to be $X_{A,0} = 0.25$. With the actual values of $r_A = 0.5$ and $r_B = 0.5$ used in Fig. 2.4, a copolymer obtained at infinitesimal conversion has, according to the *Mayo–Lewis equation*, $x_A = 0.318$. Hence, the mole fraction of monomer A decreases in the monomeric batch which, in Fig. 2.4, corresponds to a shift towards the origin of the abszissa. The composition of the related polymer can be read from the copolymerization diagram at any instant in the same way as indicated for the first portion.

2.4.1 Integration of the Mayo–Lewis Equation

The *degree of conversion*, ψ, can be derived from the *Mayo–Lewis equation*, Eq. (12.34). Precisely spoken, the left-hand side of the latter is the ratio of monomers A and B incorporated into a copolymer at infinitesimal conversion.

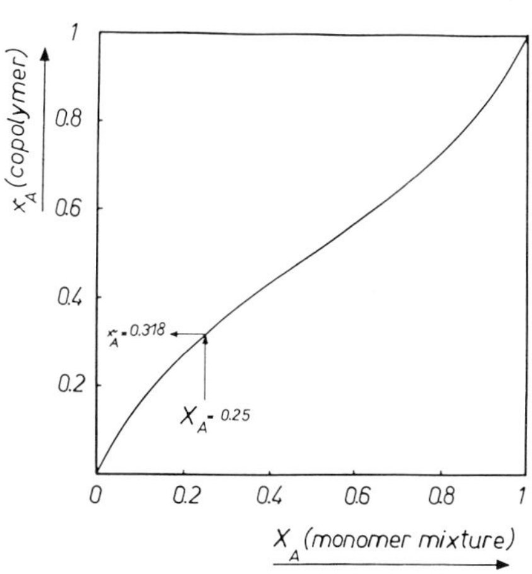

Fig. 2.4. Copolymerization diagram (graphical representation of the Mayo–Lewis equation) for $r_A = 0.5$ and $r_B = 0.5$

(In fact, the derivation yields

$$\frac{d[A]}{d[B]} = \frac{r_A[A]/[B]+1}{1+r_B[B]/[A]} \tag{2.23}$$

which is only for convenience approximated by Eq. (12.34)).

Integration of Eq. (2.23) and rearrangement gives [8, 12, 13]

$$1-\psi = \frac{X}{X_0} = \left(\frac{y}{y_0}\right)^{\alpha}\left(\frac{1-y}{1-y_0}\right)^{\beta}\left(\frac{y-X^*}{y_0-X^*}\right)^{\gamma} \tag{2.24}$$

with denotations y_0 $(=X_{A,0})$: mole fraction of A monomers in the initial mixture; $y(=X_A)$ corresponding composition of the residual batch at conversion ψ; $X_0 = X_{A,0} + X_{B,0}$; $X = X_A + X_B$; $\alpha = r_B/(1-r_B)$; $\beta = r_A/(1-r_A)$; $\gamma = (r_B r_A - 1)/[(1-r_A)+(1-r_B)]$; and $X^* = (1-r_B)/[(1-r_A)+(1-r_B)]$. (For a system with $r_A < 1$ and $r_B < 1$ or $r_A > 1$ and $r_B > 1$, X^* is the composition of the *azeotropic copolymer*.)

From the mass balance of a binary copolymerization

$$X_A(1-\psi) + \overline{x_A}\psi = X_{A,o} \tag{2.25}$$

where $\overline{x_A}$ is the average composition of copolymer, it follows:

$$\overline{x_A} = \frac{X_{A,o} - (1-\psi)X_A}{\psi} \tag{2.26}$$

Equation (2.26) combines the overall composition of a copolymer formed in polymerization up to conversion ψ with the actual composition of the monomeric mixture at this instant. Thus, the course of the reaction can be followed by the conversion dependence of the polymer as well as by the residual monomer mixture. This has been done by Stejskal et al. in a thorough investigation of the system styrene/2-methoxyethyl methacrylate [14].

The degree of conversion can be calculated from the composition of the monomeric batch by means of Eq. (2.24) and, hence, the average composition of the copolymer by means of Eq. (2.26). This way, the change in composition of the residual mixture as well as the average copolymer composition can be estimated as a function of conversion [15].

2.4.2 Distribution of Chemical Composition due to Conversion

The relative amount of copolymer with a given composition, i.e., the differential chemical composition distribution H(x), is given by the derivative of ψ vs instantaneous copolymer composition, x,

$$H(x) = \frac{1}{\psi_t}\left|\frac{d\psi}{dx}\right| \tag{2.27}$$

(In incomplete reactions, the final conversion ψ_t is a normalization factor. The derivative $d\psi/dx$ is written in its absolute value because its sign depends on the choice of component used in expressing the copolymer composition.)

Meyer and Lowry [13] derived H(x) by differentiating Eq. (2.24)

$$\frac{1}{X_o}\frac{dX}{dy} = \frac{X}{X_0}\left(\frac{\alpha}{y} - \frac{\beta}{1-y} + \frac{\gamma}{y - X^*}\right) \tag{2.28}$$

and dividing by the derivative dx/dy of Eq. (2.29). The latter is the familiar *Mayo–Lewis equation* written with mole fractions

$$x = \frac{y^2(r_A - 1) + y}{y^2(r_A + r_B - 2) - 2y(r_B - 1) + r_B} \tag{2.29}$$

where y has the meaning as in Eq. (2.24) and $x = x_A = n_A/(n_A + n_B)$ indicates the mole fraction of A units in an instantaneous copolymer polymerized from a mixture of composition y.

The more general weight form of the CCD has been presented by Molau [16] as well as by Myagchenkov and Frenkel [17]:

$$\begin{aligned} H(w) = \Bigg| &\left(\frac{W}{W_o}\right)^{\alpha}\left(\frac{1-W}{1-W_o}\right)^{\beta}\left(\frac{W - W^*}{W_o - W^*}\right)^{\gamma} \times \left(\frac{\alpha}{W} - \frac{\beta}{1-W} + \frac{\gamma}{W - W^*}\right) \\ &\times \frac{\{W^2[(r_A - 1) + q(r_B - 1)] + W(1 + q - 2qr_B) + qr_B\}^2}{W^2(r_A - 2qr_Ar_B + q^2r_B) + 2Wqr_B(r_A - q) + q^2r_B}\Bigg| \end{aligned} \tag{2.30}$$

Here, $W = W_A$ is the weight fraction of A in the monomeric mixture, $q = M_A/M_B$ the ratio of monomer molecular weight, and W^* the weight composition of an *azeotropic copolymer:*

$$W^* = \frac{q(1 - r_B)}{(1 - r_A) + q(1 - r_B)} \tag{2.31}$$

Equation (2.30) reduces to the molar CCD used by Meyer and Lowry [13], H(x), for $q = 1$, $W = X$ $(= y)$, $W_o = X_o (= y_o)$, and $W^* = X^*$.

2.4.3 Concluding Remarks

In his classic paper Stockmayer warned [5]: "It must be emphasized that the practical utility of Eq. (10) and its corollaries is severely limited, in that only the distribution for the copolymer forming at any instant is described. Since the parameters p_o, q_o, and λ will in general change as the reaction proceeds, these equations may safely be used only for copolymers of very low conversion." (In our notation: "p_o" $= x_A = x$, "q_o" $= 1 - x$, "λ" $= P_n$, "Eq. (10)" corresponds to Eq. (2.11)).

The equations describing a conversion heterogeneity should be used with similar caution. They correctly take into account the depletion of the monomeric mixture but, at high conversion, change in free-radical concentration, chain transfer to dead polymer and eventually a *gel effect* must be considered also.

Whereas initiator depletion would widen a MW distribution towards high MW values, chain transfer to dead polymer would yield combination of side chains, polymerized from the residual monomer mixture at the end of the polymerization, with backbone polymers formed at an earlier stage, i.e.

combination of parts differing in composition. The latter effect would make a chemical composition distribution narrower than curves calculated without considerations of this kind.

2.5 References

1. Benoit H (1966) Ber Bunsenges Z physik Chemie 70: 286
2. Harwood HJ, Ritchey WM (1964) J Polym Sci B Polym Letters 2: 601
3. Harwood HJ (1965) Angew Chem 77: 405, 1124
4. Simha R, Branson H (1944) J Chem Phys 12: 253
5. Stockmayer WH (1945) J Chem Phys 13: 199
6. Kuchanov SI (1978) *Metody kineticheskikh rastchotov v khimi polymerov*: Khimiya, Moskow, p 241
7. Stejskal J, Kratochvil P, Straková D (1981) Macromolecules 14: 150
8. Stejskal J, Kratochvil P (1987) Macromolecules 20: 2624
9. Schulz GV (1939) Z Phys Chem B43: 25
10. Zimm BH (1948) J Chem Phys 16: 1099
11. Tacx JCJF, Linssen HN, German AL (1988) J Polym Sci A-1, Polym Chem 26: 61
12. Mayo FR, Lewis FM (1944) J Am Chem Soc 66: 1594
13. Meyer VE, Lowry GC (1965) J Polym Sci A Gen Papers 3: 2843
14. Stejskal J, Kratochvil P, Straková D, Procházka O (1986) Macromolecules 19: 1575
15. Kruse RL (1967) J Polym Sci B Polym Letters 5: 437
16. Molau GE (1967) J Polym Sci A-1, Polym Chem 5: 401
17. Myagchenkov VA, Frenkel SY (1969) Vysokomol Soedin, Ser A-11: 2348

3 Size Exclusion Liquid Chromatography of Copolymers

Size exclusion chromatography (SEC) of polymers as an efficient method of evaluating molecular weight distributions has been dealt with in numerous reviews [1–14], some of them placing special emphasis on the separation of copolymers [2, 7, 8, 10, 12].

Let us first consider SEC where neither adsorption nor electrostatic repulsion influence separation. In ideal SEC, retention decreases with increasing value of log M, but macromolecules whose size goes beyond the exclusion limit of a column cannot be separated because they are too large to penetrate any of the pores. The opposite boundary of the separation range is the separation threshold where discrimination between molecules fails because they all are smaller than the narrowest pores of the packing and, thus, can penetrate equally. Suppliers of columns usually indicate the separation range of a packing by MW data. The limits can, for a column of given length and diameter, be also indicated by the respective values of elution volume in a chromatogram: SEC separation starts at the exclusion limit and extends to a volume due to the separation threshold. The latter corresponds to the total mobile-phase volume of a column. Retention beyond this limit signalizes adsorption. The elution volume at the exclusion limit corresponds to the *interstitial volume* of a column.

SEC is a secondary technique and the relation between log M and V_e must be established by calibration with standards whose MW has been determined by absolute methods. The calibration line is almost linear, at least with suitable combinations of columns or packings. Of course, the linear approximation cannot extend to the bounds of the separation range. Thus, a polynomial calibration is usually preferred.

SEC of copolymers is encountered with several peculiarities. A copolymer sample may contain portions of different chemical composition which effects SEC separation. This is dealt with in Sect. 3.1. Section 3.2 deals with secondary exclusion and adsorption in copolymer SEC, and Sect. 3.3 with detection problems. Finally, the effect of CCD on separation by molecular weight is discussed.

3.1 Universal Calibration

Grubisic, Rempp, and Benoit showed in 1967 that SEC retention of 8 different sets of polymers (including block and graft copolymers, PVC, PMMA, polybutadienes and poly(phenyl siloxanes)) yielded a common curve together with narrow MWD polystyrene standards when log $[\eta]M$ was plotted vs V_e [15]. (Needless to say that in a simple calibration plot, i.e. log M vs V_e, different curves were obtained with the individual sets of samples.) The eluent was THF throughout, the column a bank of four units packed with crosslinked PS. The symbol $[\eta]$ denotes the *intrinsic viscosity*.

Figure 3.1 is a photocopy of the plot log $[\eta]M$ vs V_e. (In the original publication, the diagram contains a small smut. This misprint achieved a remarkable career: it was included as a data point in some reviews of Grubisic's work and related to linear PS or comb PS, see Fig. 3.2 where it is indicated by arrows.

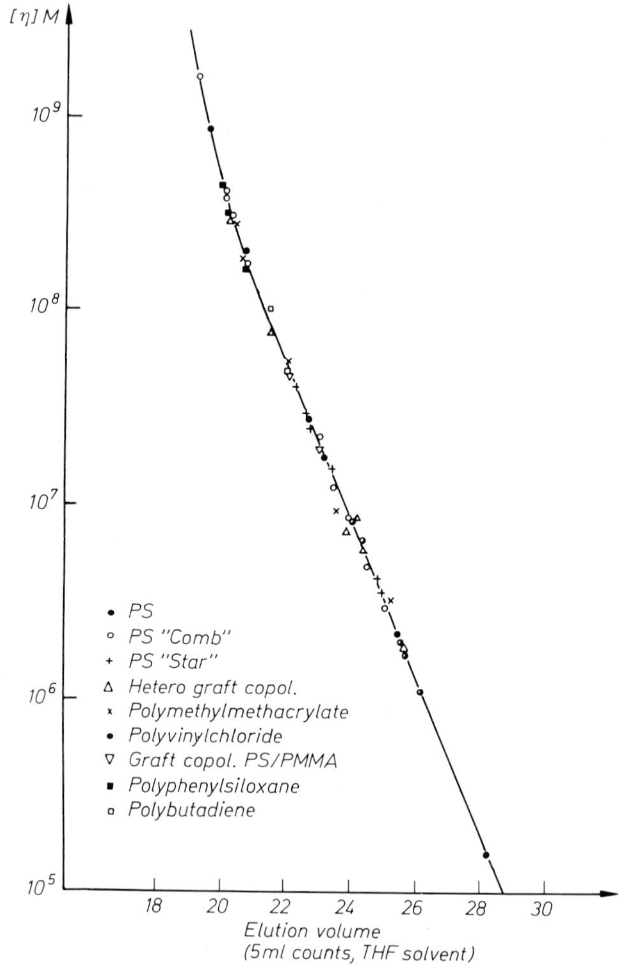

Fig. 3.1. Universal calibration. Reproduced from Ref. [15] with permission

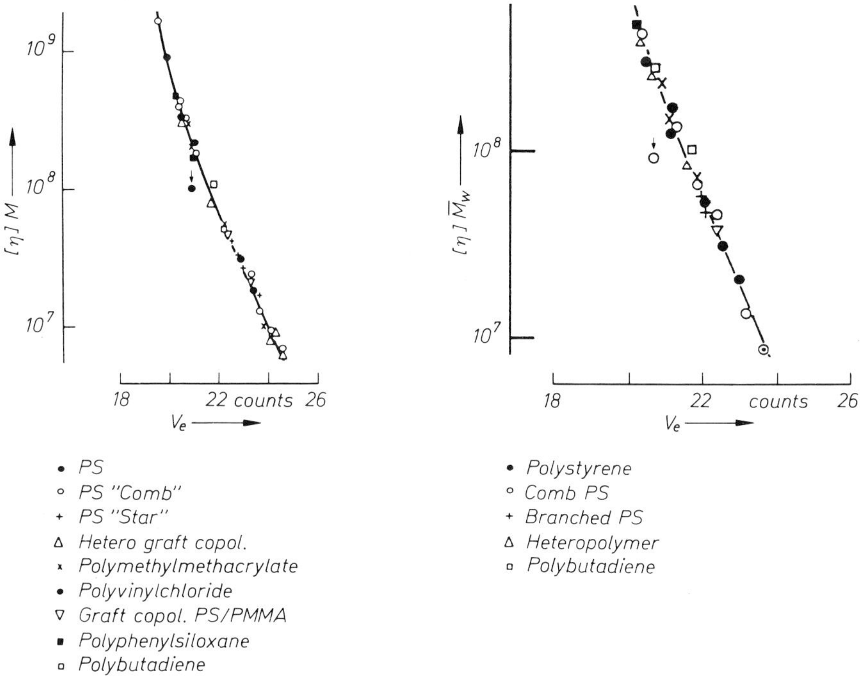

Fig. 3.2. Sections of textbook reproductions of Fig. 3.1

The product $[\eta]M$ is, through Einstein's viscosity law, related to the **hydrodynamic volume** of the particles,

$$V_h = [\eta]M/2.5 \tag{3.1}$$

The hydrodynamic volume is defined as the molar volume of impenetrable spheres which would have the same frictional properties or enhance the viscosity to the same degree as the actual polymer in solution [2].

The validity of universal calibration has been verified in numerous investigations. It holds true even with different solvents. Provided that no secondary effects are operative, one can rely on the consistency of the hydrodynamic volume with the volume that governs the entry of molecules into the pores of a SEC packing. From universal calibration it follows (for given values of elution volume):

$$[\eta]_x M_x = [\eta]_{PS} M_{PS} \tag{3.2}$$

where subscripts "X" or "PS" indicate data of an unknown polymer or of calibration standards, respectively. (The latter may be polystyrene or other suitable polymer standards.) With complex polymers and non-uniform samples, the *number average* MW has to be used in universal calibration [16]. Since intrinsic viscosity is related to MW by the **Kuhn–Mark–Houwink equation**

(KMH eq.)

$$[\eta] = K_\eta M^a \tag{3.3}$$

it follows from Eq. (3.2):

$$\log M_x = \frac{1}{1+a_x} \log \frac{K_{\eta,PS}}{K_{\eta,x}} + \frac{1+a_{PS}}{1+a_x} \log M_{PS} \tag{3.4}$$

Provided that the KMH constants $K_{\eta,x}$ and a_x are available, Eq. (3.4) enables the MW of unknown samples to be calculated from the MW data of calibrating

Table 3.1. KUHN–MARK–HOUWINK parameters in tetrahydrofuran (K in ml/g, MW range in $10^3 M$).

Polymer	°C	a	$10^2 K_\eta$	Range	Ref.
Polystyrene	25	0.694	1.622		[32]
	25	0.706	1.60	≧ 3	[33]
	25	0.713	1.25	4–7100	[34]
	25	0.717	1.17	5–867	[35]
	25	0.725	1.17	10–1000	[18]
	30	0.731	1.05	36–1000	[36]
	25	0.768	0.609		[29]
4 branches	25	0.716	0.832		[32]
6 branches	25	0.716	0.61		[32]
Amylose acetate	25	0.70	108.0	20–500	[37]
Amylose butyrate	25	0.70	111.0	20–500	[37]
Amylose propionate	25	0.61	248.0	20–500	[37]
Cellulose diacetate	25	0.688	513	60–265	[38]
Cellulose nitrate	25	1.00	25.0	95–2300	[39]
Cellulose tricarbanilate	25	0.92	0.201	38–1000	[40]
Copoly(butadiene/acrylonitrile) 33 mass% AN	25	0.646	4.56	70–470	[41]
42	25	0.833	0.094	150–1000	[41]
Poly(divinylether-*co*-maleic anhydride), methyl ester (1:2 copolymer)	30	0.50	4.895	15–543	[42]
Copoly(α-methylstyrene/acrylonitrile) 15.4 mass %AN	25	0.690	2.40		[43]
27.7	25	0.643	4.54		[43]
46.0	25	0.626	4.89		[43]
Copoly(ethylene/vinyl acetate)					
	20	0.692	1.55		[44]
	20	0.62	9.7	< 87, *linear*	[45]
	20	0.44	77.8	12–331, *br*	[45]
Copoly(isobutene/isoprene), butyl rubber	25	0.75	0.85	4–4000	[46]
Copoly(styrene/acrylonitrile, 38.3 mol% AN)	25	0.68	2.15	100–780	[47]
Copoly (S/butadiene)					
SBR (25% S, 20% vinyl)	25	0.693	4.1	24–376	[48]
SBR 1507	30	0.70	3.0	10–1000	[46]
SBR 1808	30	0.66	5.4	10–1000	[46]

(*Continued*)

Table 3.1. (*Continued*)

Polymer	°C	a	$10^2 K_\eta$	Range	Ref.
Copoly (S/butyl methacrylate)					
	25	0.800	0.370	154–1680	[29,49]
Copoly (S/ethyl methacrylate)					
	25	0.564	10.50	38–2270	[29,49]
Copoly (S/methyl acrylate)					
45–75 mol% MA	25	0.799	0.44	70–600	[50]
		0.799	0.37		[51]
Copoly (S/methyl methacrylate)					
23% MMA	25	0.754	0.848	465	[52]
36	25	0.704	1.853	438	[52]
48	25	0.722	1.229	506	[52]
56	25	0.733	1.169	499	[52]
68	25	0.803	0.390	659	[52]
32.7 mass% MMA		0.719	1.28		[53]
34.3		0.7185	1.29		[53]
44.4		0.718	1.33		[53]
44.5		0.717	1.36		[53]
58.5		0.738	1.00		[53]
67.5		0.797	0.49		[53]
50 (alternating)	25	0.76	0.775	71–2330	[54]
50 (block copolymer)		0.76	0.641	32–144	[54]
Copoly (S/monoethyl maleate)					
	25	0.695	0.75	200–1800	[55]
Copoly (S/octyl methacrylate)					
	25	0.643	2.394	194–2000	[29,49]
Copoly (vinyl acetate/ vinyl chloride)					
87–90 mass% VC	25	0.611	6.72	30–250	[56]
71.9–94	25	0.72	1.8	18–250	[57]
90.3	25	0.742	1.64	18–88	[58]
85	25	0.746	1.81	6–102	[59]
Polyarylate	25	0.575	7.25	7–234	[60]
	25	0.61	6.17	18–86	[60]
Polybutadiene (*cis*/*trans* ≈ 0.8)					
8% vinyl	25	0.693	4.57	81–1146	[48]
28% vinyl	25	0.693	4.51	19–202	[48]
52% vinyl	25	0.693	4.28	19–183	[48]
73% vinyl	25	0.693	4.03	19–191	[48]
Polybutadiene (57% *trans*, 36% *cis*, 7% 1,2 –)	30	0.74	2.56	11–571	[61]
Poly-1,2-butadiene	ns	0.85	0.601	23–880	[62]
Poly-1,4-butadiene	25	0.776	1.6	43–282	[34]
Poly (butyl methacrylate)	25	0.758	0.503	99–1330	[29,49]
Polycarbonate	25	0.67	4.9	7–77	[63]
	25	0.673	4.4		[64]
	ns	0.70	3.99	10–270	[65]
Poly (ethyl methacrylate)	25	0.679	1.549	270–2050	[29,49]
Polyisoprene	25	0.735	1.77	40–500	[48]
natural rubber	25	0.79	1.09	10–1000	[66]
Poly (*n*-lauryl methacrylate)					
	25	0.69	0.73	32–2220	[67]
Poly (methyl acrylate)	25	0.82	0.388	110–600	[68]

(*Continued*)

Table 3.1. (*Continued*)

Polymer	°C	a	$10^2 K_\eta$	Range	Ref.
Poly (methyl methacrylate)	25	0.406	21.1	< 31	[69]
	25	0.697	1.04	≧ 31	[69]
	25	0.677	1.48		[70]
	25	0.66	1.99	40–240	[71]
	25	0.69	1.28	150–1200	[39]
		0.69	1.78	24–94	[72]
		0.69	1.795		[72]
	23	0.72	0.93	170–1300	[15]
	25	0.72	0.868	210–1075	[73]
		0.746	0.73	700–3000	[74]
Poly (methoxyethyl methacrylate)	25	0.71	0.757	45–206	[75]
Poly (octadecyl methacrylate)	30	0.75	0.25	230–1670	[76]
Poly (octadecyl vinylether)	30	0.35	22.4	9–110	[76]
Poly (octyl methacrylate)	25	0.560	5.556	194–2090	[29, 49]
Poly (oxypropylene), isotact.					
	20	0.62	5.5	0.5–3.3	[77]
Poly (stearyl methacrylate)	30	0.67	0.9	15–935	[61]
	30	0.708	0.579		[70]
Poly (*t*-butyl styrene)	30	0.70_0	1.0_4	27–175	[78]
Poly (vinyl acetate)	25	0.70	1.6	55–500	[79]
	25	0.791	0.51	302–625	[34]
Poly (vinyl bromide)	25	0.64	1.59	20–100	[80]
Poly (vinyl carbazol)	25	0.65	1.44	10–45	[81]
Poly (vinyl chloride)	25	0.69	4.98		[82]
	25	0.700	4.48		[29, 49]
	25	0.766	1.63	20–170	[83]
	25	0.77	1.35		[84]
	25	0.77	1.50	10–120	[85]
	25	0.77	1.60	10–1000	[18]

polymers delivered at the same elution volume ($V_{e,x} = V_{e,PS}$). Tables 3.1 and 3.2 provide KMH data measured in THF or DMF, respectively. For data in other solvents see, e.g.[1]

If $K_{\eta,x}$ and a_x are not available, the wanted $(MW)_x$ can be derived from measurements performed with an SEC instrument equipped with an additional viscosity detector which yields $[\eta]_{x,i}$ for incremented values of elution volume, $V_{e,i}$. Hence, the measured curve $[\eta]_{x,i} M_{x,i}$ vs $V_{e,i}$ (equivalent to $[\eta]_{PS,i} M_{PS,i}$ vs $V_{e,i}$) can be transformed into the desired $M_{x,i}$ vs $V_{e,i}$ result [18, 19].

Direct monitoring of the MW of a solute by, e.g. a light-scattering detector can be taken into account as well. Instruments with laser light-sources are commercially available which measure scattered light either at a low angle [20, 21] or simultaneously at several fixed angles between 7° and 170° [22]. Since the refractive index increment must be known at any composition of the eluting

[1] The compilation provided by Kurata et al. [17]. Additional data for PS and some special polymers can be found in ASTM Standard D-3593-80

Table 3.2. KUHN-MARK-HOUWINK parameters in dimethyl formamide (K in ml/g, MW range in 10^3 M).

Polymer	°C	a	$10^2 K_\eta$	Range	Ref.
Polystyrene	25	0.6	3.32	168–465	[86]
	30	0.688	1.05	237–624	[87]
	40	0.697	0.94		[87]
	50	0.716	0.71		[87]
	60	0.728	0.59		[87]
Copoly (acrylic acid/ methyl methacrylate) 42 mol% MMA	ns	0.89	3.0	34–855	[88]
Copoly (acrylonitrile/methyl acrylate)	20	0.79	1.79	20–210	[89]
Copoly (acrylonitrile/ methyl methacrylate) 52% MMA	30	0.65	4.61	667	[90]
Copoly (styrene/acrylonitrile) 13.0 mass% AN	20	0.71	1.45	16–1070	[27]
24.4	20	0.71	1.665		[27]
52.0	20	0.72	2.65	15–800	[27]
27.4% AN	30	0.74	1.20	292	[91]
38.9%	30	0.73	1.62	492	[91]
47.5%	30	0.73	1.72	550	[91]
Copoly (S/MMA)	30	0.712	1.01	372–1113	[87]
	40	0.725	0.88		[87]
	50	0.728	0.84		[87]
	60	0.733	0.76		[87]
Polyacrylonitrile	20	0.66	25.0	9–69	[92]
	20	0.725	3.25		[27]
	25	0.73	5.74	168–352	[86]
	25	0.75	2.33	30–265	[93]
	25	0.81	1.66	48–270	[94]
non-fractionated	25	0.681	8.511	7.5–20	[95]
fractionated	25	0.792	3.981		[95]
	30	0.76	2.09	58–295	[96]
	35	0.76	2.78	28–575	[97]
Poly (ethylene oxide)	25	0.73	2.4		[98]
Poly (methacrylonitrile)	29.2	0.503	30.6	6–80	[99]
Poly (methyl methacrylate)	25	0.7	9.1	410–465	[86]
	30	0.65	1.70	82–2904	[87]
	45	0.657	1.62		[87]
	60	0.667	1.58		[87]
	30	0.65	1.70	710	[90]
Poly (2-methyl-5-vinyl-pyridine)	25	0.76	1.30	40–400	[100]
Poly (2-vinyl-pyridine)	25	0.67	1.47	30–90	[101]
Polyvinylpyrrolidone	25	0.55	5.49	70–920	[102]

copolymer, the previously discussed use of a viscosity detector is more generally applicable to complex polymers with broad distributions.

In order to check whether or not a copolymer/solvent-system adheres to the principle of universal calibration, the elution curve of a sample can be converted point by point into $[\eta]_i M_i$ values by using an $[\eta]_i M_i$ calibration obtained with suitable standards. Multiplication of these incremental $[\eta]_i M_i$ data by the corresponding concentration and summation should give the $[\eta]M$ value which also results from direct molecular-weight measurement of the whole sample (preferably by a method which gives the number average MW [16]) and its intrinsic viscosity. The procedure should be performed on various samples of the system; it neither requires chemically homogeneous specimens nor the knowledge of KMH relations of the system under investigaion, but care must be taken with respect to c_i measurement (see Sect. 3.3).

3.2 Secondary Exclusion and Adsorption Effects

Adherence to universal calibration can be regarded as confirmation of an ideal size-exclusion mechanism. Real SEC is influenced by additional effects. Exclusion chromatography is based upon the restricted distribution of solutes (K_{excl} in Fig. 3.3) between the *interstitial volume* of a column and the *pore volume* while interaction chromatography (adsorption chromatography, reversed phase chromatography, ect.) is due to partition between mobile and stationary phase (K_{ads} in Fig. 3.3). The distribution coefficients K_{excl} and K_{ads} have different origins. Discrimination between these distribution systems must not be ignored. Size exclusion chromatography yields elution within the SEC separation range, i.e. between exclusion limit and separation threshold. Within the separation

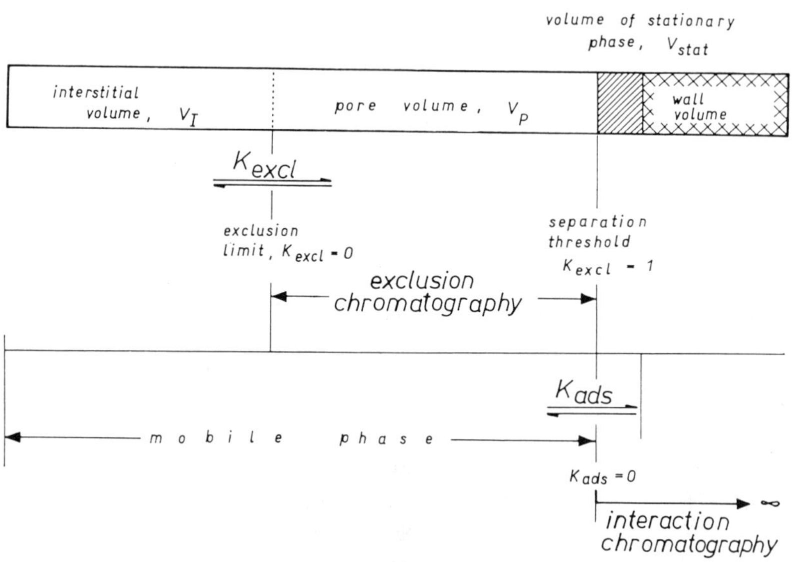

Fig. 3.3. Schematic representation of elution ranges in real chromatography of polymers on porous packings. In ideal interaction chromatography, the mobile-phase volume is $V_{mob} = V_I + V_P$ and retention is due to partition between V_{mob} and V_{stat}; a component with $K_{ads} = 0$ is eluted at V_{mob}. In ideal exclusion chromatography, the stationary phase can be considered part of the wall because it is impermeable to the solute; retention is due to distribution between V_I and V_P; a component with $K_{excl} = K_{SEC}K_{el} = 0$ is eluted at V_I

range, the elution volume of a solute is larger than the interstitial volume but smaller than the mobile phase volume of a column. Interaction chromatography (without permeability restrictions) is characterized by elution volumes larger than the mobile phase volume of a column. A schematic representation is given in Fig. 3.3.

Provided that the flow rate is sufficiently small and does not disturb the equilibrium, real SEC can be considered as follows. The distribution of solute molecules between interstitial and pore volume is entropy-driven. Due to the gain in entropy, the process is, of course, linked to a small but definite decrease in Gibbs Free Energy. With flexible polymer molecules, the distribution can be described also with reference to the conformational entropy of the macromolecules:

$$\Delta G = \Delta H - T\Delta S = -\mathrm{R}T \ln K_{\mathrm{excl}} \tag{3.5}$$

where ΔG is the change in Gibbs Free Energy, ΔH the change in enthalpy, and ΔS the conformational entropy. Since ΔH is zero in ideal SEC, a real distribution coefficient K_{excl} can be formally subdivided into

$$K_{\mathrm{SEC}} = \exp(\Delta S/\mathrm{R}) \tag{3.6}$$

and

$$K_{\mathrm{el}} = \exp(-\Delta H/\mathrm{R}T) \tag{3.7}$$

where K_{SEC} represents the distribution coefficient of an ideal size exclusion mechanism. The insertion of a large molecule into a pore diminishes the degree of freedom of the molecule, hence, $\Delta S < 0$ and $K_{\mathrm{SEC}} < 1$. The maximum value, $K_{\mathrm{SEC}} = 1$, is related to zero change in conformational entropy. K_{el} is effected by enthalpic interactions; these are in the first instance electrostatic repulsion forces, which on diffusion of a particle into a pore would cause a positive change in ΔH and, thus, $K_{\mathrm{el}} < 1$.

Ideal SEC occurs on condition that $K_{\mathrm{el}} = 1$ (or $\Delta H = 0$, vide supra). The additional effect of electrostatic repulsion is referred to as secondary exclusion. The product $K_{\mathrm{SEC}}\ K_{\mathrm{el}}$ governs the real exclusion of a solute.

Formally, Eq. (3.7) should apply also for $\Delta H < 0$. As a consequence of a negative change in enthalpy, the exclusion of a given polymer would decrease. Indeed, the penetration of large molecules into narrow pores is under discussion, but attractive forces will eventually cause adsorption of the solute, which is a partition between mobile and stationary phase, see Fig. 3.3. The latter effect must be taken into account by an additive term, K_{ads}, in the retention equation:

$$V_{\mathrm{e}} = V_{\mathrm{I}} + V_{\mathrm{P}}(K_{\mathrm{SEC}}K_{\mathrm{el}}) + K_{\mathrm{ads}}V_{\mathrm{stat}} \tag{3.8}$$

Assume $K_{\mathrm{el}} = 1$ (i.e. $\Delta H = 0$, no electrostatic repulsion) and $K_{\mathrm{ads}} = 0$. Then, Eq. (3.8) reduces to

$$V_{\mathrm{e}} = V_{\mathrm{I}} + V_{\mathrm{P}}K_{\mathrm{SEC}} \tag{3.9}$$

describing SEC which is properly reflected by universal calibration. Note that electrostatic repulsion ($K_{el} < 1$) would cause exclusion even for $K_{SEC} = 1$.

For $K_{SEC} = K_{el} = 1$,

$$V_e = V_I + V_P + K_{ads} V_{stat} \tag{3.10}$$

describes ideal adsorption chromatography without accessibility restrictions. For large molecules on small pore packings (i.e. $K_{SEC} = 0$),

$$V_e = V_I + K_{ads} V_{stat} \tag{3.11}$$

characterizes adsorption chromatography on the outer surface of a packing. In this case, an unretained solute is eluted at the *interstitial volume.* Many of the separations reported in Chap. 9 were performed under the conditions described by Eq. (3.11).

Real SEC is often mixed-mode chromatography, hopefully with predominance of K_{SEC}. With homopolymers, the extra effects corresponding to K_{el} and K_{ads} cause the calibration curve to be shifted from the position predicted from universal calibration towards higher or lower values of V_e, but do not falsify the shape of elution curves.

With chemically heterogeneous samples, effects are more dramatic because secondary exclusion and adsorption act differently on molecules of different composition. This causes the direction of separation to diverge from the direction set by hydrodynamic volume [23]. The elution curves are distorted and will no longer reflect the size distribution of the samples injected.

Hence, copolymer SEC requires conditions which suppress secondary exclusion and adsorption. Electrostatic repulsion is likely to occur at high values of dielectric constant, i.e., preferably in aqueous eluents. Adjustment of ionic strength and pH, or addition of salt or surfactants have found to counteract electrostatic repulsion and adsorption in aqueous SEC. Adsorption in purely organic (nonaqueous) systems must be suppressed by strong eluents and inert packings.

3.3 Detector Response

The most widely used detection principles are measurements of UV absorption or of refractive index. They perform well with homopolymers, provided that both eluent and solute have suitable optical properties. With copolymers, the chemically differing units of the constituting monomers will in general differ in absorbance and refractive index increment. Thus, the detector response to a *bipolymer* is generally given by

$$R_i = c_i[w_A R_A + (1 - w_A) R_B] \tag{3.12}$$

where R_i is the signal caused by a sample portion of concentration c_i in the detector cell, w_A the weight fraction of A units in the sample, and R_A and R_B the contributions of A or B units to the signal. The composition-dependent curve monitored in copolymer elution can be transformed into a plot of polymer concentration vs elution volume when all parameters of Eq. (3.12) are known.

Additional care must be taken if the contributions R_A or R_B depend on sequence length. The influence of neighbouring units on the UV absorbance of styrene has been found with copolymers S/AN [24, 25] and S/MMA [26].

On the other hand, the use of two (or more) detectors enables the average composition of the copolymer in each portion of the eluate to be derived from the signal ratio. Sometimes, this dual (or multi-) detection SEC is considered a method for the evaluation of composition heterogeneity of a copolymer. If the ratio changes in the course of the elution, the sample is certainly heterogeneous but, since mixtures of widely differing composition can yield one and the same average, a constant ratio does not say whether or not a sample is homogeneous.

3.4 Fractionation by Hydrodynamic Volume

The prediction of SEC elution of a copolymer with a CCD requires the knowledge of KMH parameters as a function of chemical composition. Unfortunately, information of that kind is published for only a few systems. Usually, neither K_η nor a vary linearly with composition, but the deviation from linearity is often opposite for K_η and a which, by the combination of both values according to the KMH equation, considerably reduces the effect on intrinsic viscosity. Figs. 3.4-a to 3.4-f are graphical representations of published data, which show the composition effect on K_η, a, and $[\eta]$ (calculated for polymers with degree of polymerization, $P = 1000$).

With copolymers it is more advantageous to compare samples of equal chain length instead of equal MW, especially if the structural units differ substantially in MW. This has been stated by Lange and Baumann as early as 1970 [27]. Figure 3.5 shows that $[\eta]$ of vinyl polymers (calculated for $M = 100{,}000$) decreases with increasing MW of the repeat units, but it must be emphasized that a related plot for constant P does not at all yield constant $[\eta]$ for different polymers. Figure 3.6 shows the composition effect on $[\eta]$ for S/AN copolymers, calculated from KMH data of Ref. [27] for $P \Rightarrow$ const or $M =$ const. The effect is larger in the latter case, but is substantial also for $P =$ const. Thus, the evaluation of SEC measurements [28] with the help of equations exclusively taking into account the effect of repeat-unit MW on $[\eta]$ can only be used with caution and must not be generalized.

In several cases, the exponent a is higher at intermediate composition than at the homopolymer edges of the plots. The maximum bears similarity to the increased solubility of copolymers and can be understood as the consequence of repulsive interactions between structurally differing units.

3.4.1 Intrinsic Viscosity Varying Linearly with Copolymer Composition

A linear dependence of $[\eta]$ on composition can, for a certain value of MW, result by chance even from KMH constants varying nonlinearly. Linearity for any MW data requires rather special dependence of both K_η and a on composition.

Retention in SEC varies with log $[\eta]M$. Thus, a linear dependence of $[\eta]$ on copolymer composition does not imply linearity between composition and V_h. In

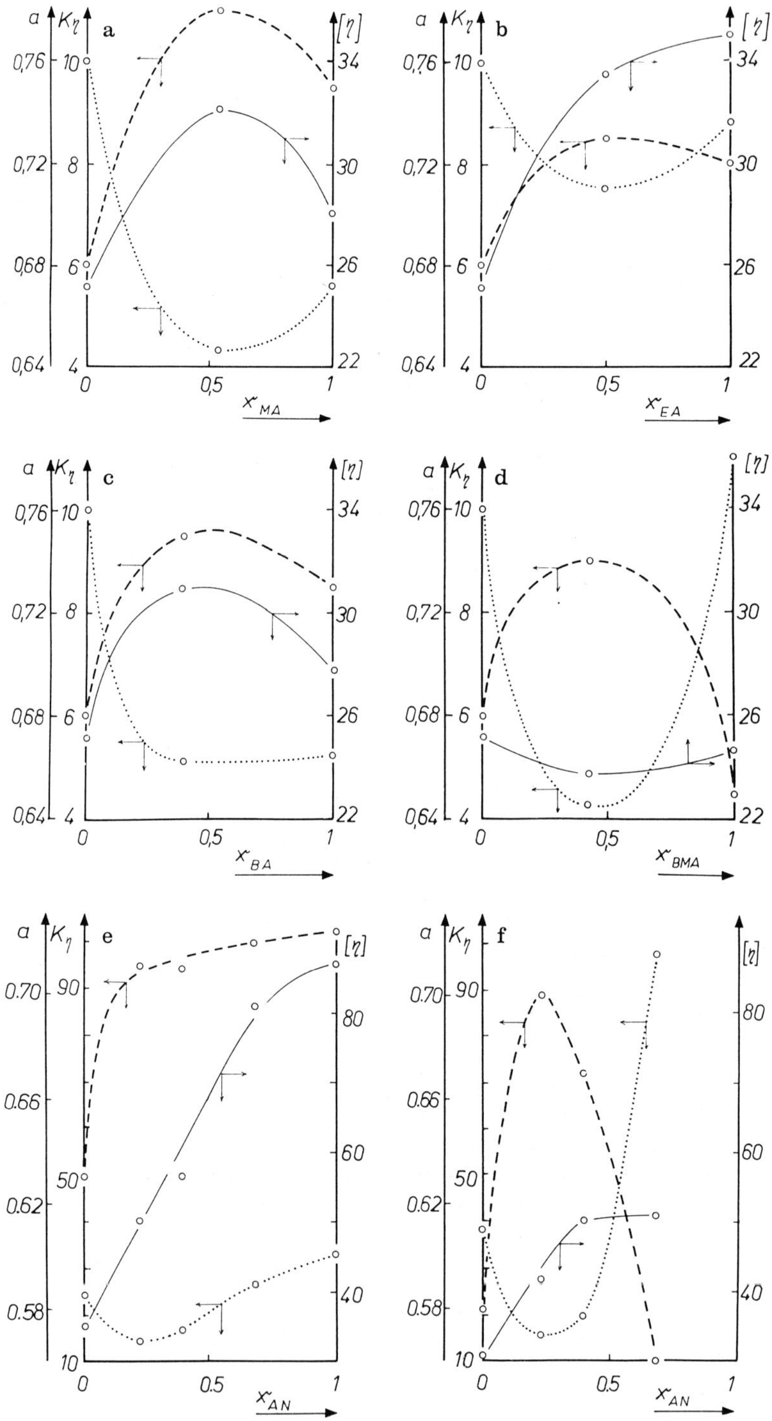
a
a
K_η
$[\eta]$
0,76
0,72
0,68
0,64
10
8
6
4
34
30
26
22
0
0,5
1
x_{MA}
b
x_{EA}
c
x_{BA}
d
x_{BMA}
e
0.70
0.66
0.62
0.58
90
50
10
80
60
40
0.5
x_{AN}
f

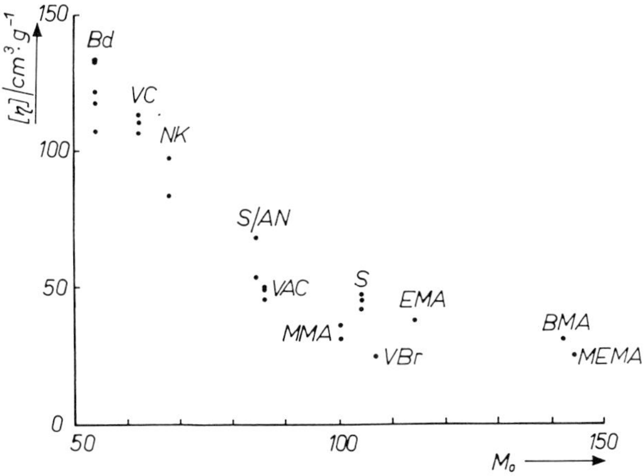

Fig. 3.5. Intrinsic viscosity of vinyl polymers vs MW of the respective repeat unit. Calculated from KMH data of Table 3.1 for $M = 100{,}000$. (Local abbreviations: VC vinyl chloride, NK natural rubber, VBr vinyl bromide.)

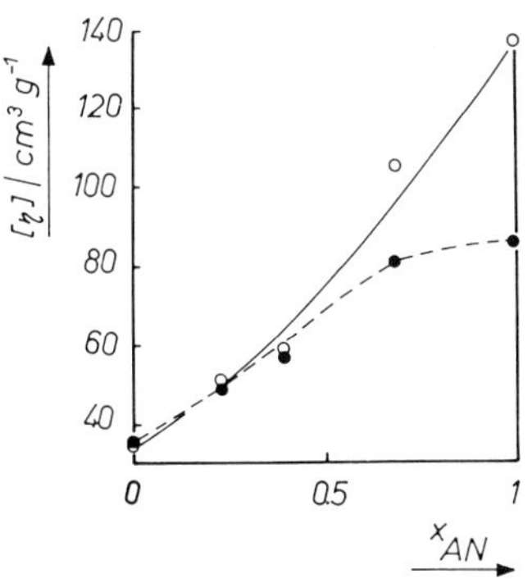

Fig. 3.6. Effect of copolymer composition on intrinsic viscosity of S/AN in DMF, 20 °C, calculated for $M = 100{,}000$ (○, *full line*) or $P = 1000$ (●, *dashed line*). KMH data taken from Ref. [27], see Fig. 3.4e

the rare case of $[\eta]$ being independent of composition, V_h would also be independent of composition. In a system of this kind, fractions taken at increasing values of V_e contain molecules of decreasing size. Each fraction comprises molecules in a certain range of MW and CC, which corresponds to the volume of the fraction. The CCD within each fraction is determined exclusively by the MW/CCD of the copolymer under investigation.

◀

Fig. 3.4 a–f. Intrinsic viscosity and KMH data of bipolymer systems. Figures a-d: acrylate polymers in acetone, 20 °C, (**a**) MMA/MA, (**b**): MMA/EA, (**c**): MMA/BA, (**d**): MMA/BMA, abscissa: mole fraction of the indicated monomer, KMH data from Ref. [31]. Figures (**e**) and (**f**): S/AN in DMF at 20 °C (**e**) or MEK at 30 °C (**f**), KMH data from Ref. [27]. *Continuous curves*: $[\eta]$ calculated for polymers of equal length ($P = 1000$) from KMH factor K_η (given in 10^{-3} ml/g, *dotted lines*) and exponent *a* (*dashed lines*)

In all other instances, i.e., where parent homopolymers of identical MW differ in [η], the CCD in the fractions is dependent upon sample MW/CCD and separation characteristic of the system.

3.4.2 Nonlinear Composition Dependence of Intrinsic Viscosity without Extreme Values

If all [η] data are within the limits set by the related homopolymers, a nonlinear composition dependence will not impair SEC separation by molecular size. In principle, the effect of a nonlinear composition dependence on [η] as well as the curvature of the V_h vs CC plot due to logarithmic transformation can be treated in the same manner. Of course, in both instances the inclination of the characteristics V_h vs CC influence the result of SEC fractionation. A set of lines which rise on transition from 0–100% B units in the polymer (see Fig. 3.7) has the consequence that the first fractions have higher MW and higher content in B than the subsequent ones. Note that this effect is caused only by separation characteristics and bears no relation to any correlation between MWD and CCD.

In general, similar qualifications are required as discussed in the previous section. The variation in slope of the $[\eta]M$ vs CC curves causes the separation efficiency to vary in the course of a fractionation, but this can be compensated by intentionally varying the volume of subsequent fractions, if necessary.

3.4.3 Nonlinear Composition Dependence with Extreme Values

In several cases, a composition effect with a maximum in [η] has been found, e.g. with S/EMA copolymers in THF [29]. Since this extremum is very pronounced, an examination of this system could show what may happen at worst, see Fig. 3.8.

The maximum in [η] is caused by the high value of K_η measured with a copolymer of almost equimolecular composition (49.7 wt% EMA). The exponent a has a slight minimum in the same range of composition which diminishes the effect of K_η on [η] to some degree. When, as in the system under consideration,

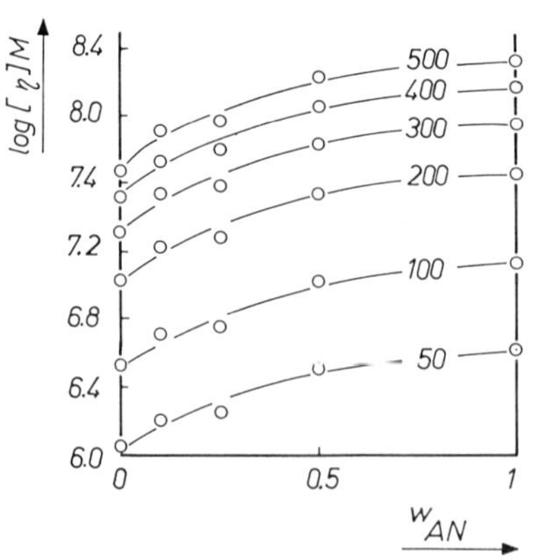

Fig. 3.7. Effect of weight fraction acrylonitrile and MW on hydrodynamic volume of S/AN copolymers in DMF at 20 °C, molecular weight (10^{-3} M) indicated, KMH constants from Table 3.2, Ref. [27]

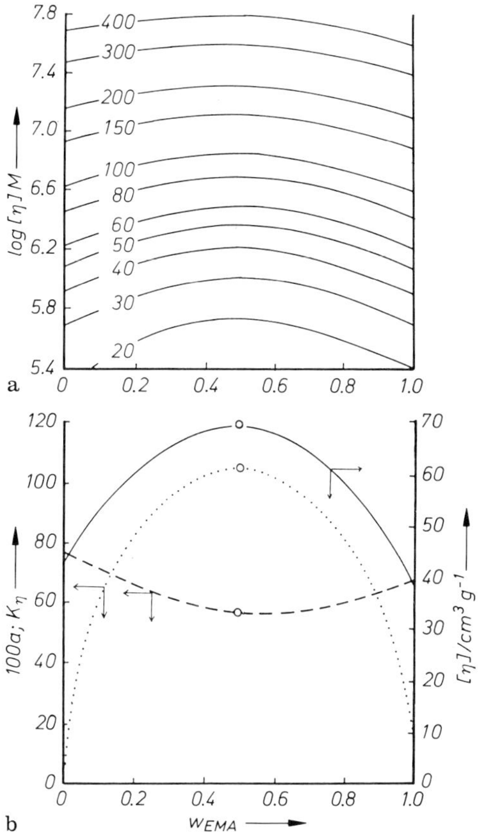

Fig. 3.8 a, b. Hydrodynamic volume (**a**) and Kuhn–Mark–Houwink constants (**b**) of copoly(styrene/ethyl methacrylate) in THF at 25 °C, molecular weight ($M \times 10^{-3}$) indicated. Figure 3.8b: *full line*: intrinsic viscosity, calculated for $M = 100{,}000$ from K_η (*dotted curve*) and *a* (*dashed curve*); abscissa: mass fraction EMA. Data from Ref. [29]

the extremum in $[\eta]$ is dominated by a maximum in K_η, the effect of the latter on $[\eta]$ decreases with increasing MW. (The opposite would be true for extreme values dominated by a maximum in exponent *a*, of course.)

Since SEC fractionation is governed by the logarithm of V_h, the influence of any extreme value of $[\eta]$ on the separation is reduced the more the molecular weight increases. This is demonstrated in Fig. 3.8a where the curvature in the upper part is smaller than in the lower part.

Thus, for S/EMA system, one has to expect SEC fractions which comprise high-MW portions of intermediate composition with portions of lower MW from the wings of the CC distribution. The effect is less pronounced with high-MW fractions than with low-MW ones and, in the whole, not dramatic.

3.4.4 Concluding Remarks

If SEC fractionation is the first stage in cross fractionation, information on the composition dependence of hydrodynamic volume is indispensable for the proper interpretation of the final results. Unfortunately, the KMH data available

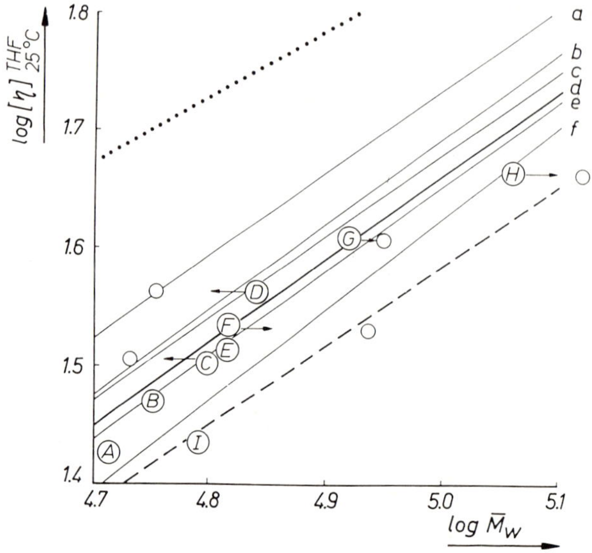

Fig. 3.9. Intrinsic viscosity vs molecular weight of copoly(styrene/ethyl methacrylate) specimens A–I and graphic representation of Kuhn–Mark–Houwink data for polystyrene (lines *a–f*), poly(ethyl methacrylate) homopolymer ($[\eta] = 1.549 \times 10^{-2} M^{0.679}$ [29], *dashed line*) and a S/EMA sample of 49.7 wt% EMA ([29], *dotted line*). All data in THF, 25 °C. *Circles* with the indications A–I show intrinsic viscosity data plotted vs MW from SEC elution with PS calibration. The *empty circles* show $[\eta]$ plotted vs M_w measured by light scattering of the sample. KMH data of PS from Refs. [32]: line *a*, [18]: *b*, [33]: *c*, [34]: *d*, [35]: *e*, [29]: *f*. By courtesy of Springer-Verlag from Ref. [30]

differ surprisingly even for systems which are known to be reliable. This can be seen in Fig. 3.9, where the straight lines a–f show the course of several published $[\eta]$ vs M relations for PS in THF. Figure 3.9 was used for considering the chance of prefractionating S/EMA copolymers by SEC [30]. The dotted line indicates the relation $[\eta] = 0.105\ M^{0.564}$ published for an equimolar S/EMA copolymer [29]; the dashed line is the KMH characteristic of PEMA homopolymer [29].

The samples investigated [30] are indicated by capital letters A – I. The EMA content was in the range 4.7% (sample "A") – 92.5 mass% (sample "I"). (For data, see Table 9.5). The circles with the sample code indicate the measured intrinsic viscosity vs SEC molecular weight of the respective sample. The corresponding characteristic of PS is line "f" [29].

The intrinsic viscosities of all samples (besides I with 92.5% EMA) lay above the characteristic for PEMA homopolymer in a reasonable distribution around the PS line "f", but did not reach by far the high viscosity predicted by the dotted line. On base of this finding it was decided "that SEC can be used for prefractionation by MW of S/EMA specimens" [30].

The example shows that MW evaluation of copolymer SEC curves by means of KMH data is not a triviality. Even if KMH data are available in the desired range of composition, full correspondence is required between the samples under investigation and the copolymers used in the evaluation of the KMH data. Solvent quality is also an important factor. THF can cause problems in this respect because it is hygroscopic; traces of water will have a tremendous effect on solution properties of hydrophobic polymers. Furthermore, SEC evaluation by means of KMH data requires the knowledge of the actual composition of the eluting polymer. This information can be gained from the signal of an additional composition-sensitive detector (e.g. UV or IR detectors) or from the result of subsequent HPLC analysis of the SEC fractions. If HPLC elution is effected by MW, iterative calculations are required. Thus, the most favourable choice in copolymer SEC is the use of a complementary viscosity detector [18, 19], see Sect. 3.1.

Finally, we should consider again the influence of composition dependence in SEC separation (and, correspondingly, of MW effect in gradient HPLC) on the accuracy of cross-fractionation results. With present knowledge of copolymer properties, we can assume that SEC will predominantly yield separation

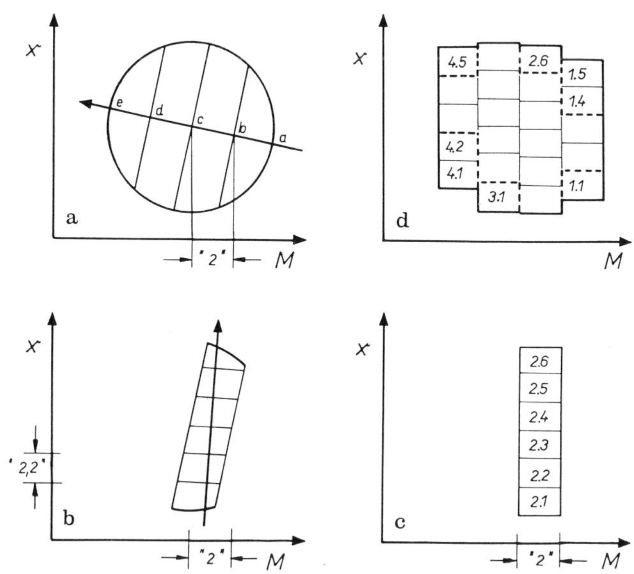

Fig. 3.10 a–d. Dependence of cross-fractionation results on the direction of separation (schematically). (**a**) SEC fractionation of a copolymer sample. The slope of the arrow *ae* indicates an effect of composition on SEC separation, which yields fractions whose boundary-lines are not perpendicular to the MW axis. With a sample of unknown MW/CCD, fraction "2" is, through the calibration of SEC, assumed to be within the limits M_c and M_b. (**b**) Gradient HPLC with a non-zero MW effect: direction of separation shown by the *arrow*. By calibration, x_i values are co-ordinated to the composition range of each HPLC subfraction (indicated for subfraction "2.2"). (**c**) Each subfraction is, by SEC and HPLC calibration, represented by a rectangle. A total SEC fraction is represented by a *column of rectangles* (indicated for fraction "2"). (**d**) An image of the MW/CC distribution of the whole copolymer is assembled by co-ordination of all pieces of information. (The portions whose numbers are indicated contain less material than the others, i.e. the *dashed line* is a first estimate of a contour line)

according to MW, see Fig. 3.8. Hence, the direction of SEC separation can usually be indicated in a CC–MW–plane by an arrow with a slight inclination, see Fig. 3.10a. The effect of MW on separation in gradient HPLC is small as well. Thus, the direction of separation by gradient HPLC can be indicated by an arrow almost parallel to the x direction, see Fig. 3.10b. Figure 3.10d shows schematically the information available by chromatographic cross-fractionation with usual calibration, i.e. without additional measurement of certain properties of subfractions. The rather crude approximation of the results to the assumed MW/CC distribution 3.10a is mainly due to separation into only four SEC fractions and five or six subfractions each. In real experiments described in Chap. 10, gradient HPLC was continuously monitored on about ten SEC fractions. With increasing number of fractions, the result of CCF will better and better reflect the true distribution of a sample. Restricted deviation from ideality provided, the influence of separation directions will become negligible at sufficiently high number of fractions.

3.5 References

1. Tung LH, Moore JC (1977) In: Tung LH (ed) Fractionation of synthetic polymers, Marcel Dekker, New York
2. Harmon DJ (1978) In: Epton R (ed) Chromatography of synthetic and biological polymers, vol 1, Horwood E, Chichester, p 137
3. Yau WW, Kirkland JJ, Bly DD (1979) Modern size exclusion liquid chromatography, Wiley-Interscience, New York
4. Moore JC (1981) In: Cazes J (ed) Liquid chromatography of polymers and related materials, III, Chromatographic science series, vol 19, Marcel Dekker, New York
5. Janča J (1981) In: Giddings JC, Grushka E, Cazes J, Brown PR (eds) Advances in chromatography, vol 19, Marcel Dekker, New York
6. Glöckner (1982) Polymercharakterisierung durch Flussigkeitschromatographie, Dr Hüthig, Verlag, Heidelberg
7. Mori S (1983) In: Giddings JC, Grushka E, Cazes J, Brown PR (eds) Advances in Chromatography, vol 21, Marcel Dekker, New York, p 187
8. Garcia-Rubio H, MacGregor JF, Hamielec AE (1983) In: Crawer CD (ed) Polymer characterization: Spectroscopic, chromatographic, and physical instrumental methods, Advances in chemistry series, No 203, Washington, p 312
9. Janca J (ed) (1984) Steric exclusion liquid chromatography of polymers, Chromatographic Science Series, vol 25, Marcel Dekker, New York
10. Quivoron C (1984) In: Janča J (ed) Steric exclusion liquid chromatography of polymers, Chromatographic Science Series, vol 25, Marcel Dekker, New York, p 213
11. Provder T (ed) (1984) Size exclusion chromatography: Methodology and characterization of polymers and related materials, ACS Symposium Series, vol 245, Washington
12. Glöckner G (1987) Polymer characterisation by liquid chromatography, J Chromatography Library, vol 34, Elsevier, Amsterdam, p 213
13. Dubin PL (ed) (1988) Aqueous size-exclusion chromatography, J Chromatography Library, vol 40, Elsevier, Amsterdam
14. Dawkins JW (1989) In: Booth C, Price C (eds) Comprehensive polymer science, vol 1, Pergamon, Oxford, p 231
15. Grubisic Z, Rempp P, Benoit H (1967) J Polym Sci B, Polym Letters 5: 753
16. Hamielec AE, Ouano AC (1978) J Liquid Chromatog 1: 111
17. Kurata M, Tsunashima Y, Iwama M, Kamada K (1975) In: J Brandrup, Immergut EH (eds) Polymer handbook, 2nd ed Wiley, New York, Chap. IV, pp 1–33, 53–60
18. Goedhardt D, Opschoor A (1970) J Polym Sci A2, Polym Phys 8: 1227
19. Grubisic-Gallot Z, Picot M, Gramin Ph, Benoit H (1972) J Appl Polym Sci 16: 2931
20. Kaye W, Havlik A (1973) J Appl Opt 12: 541

21. CHROMATIX application note LS-2 (10 MO 177). Determination of molecular-weight-distributions by combining low angle light scattering with gel permeation chromatography
22. DAWN(R) MODEL F, Wyatt Technology, Santa Barbara, CA, USA
23. Meira GR, Garcia-Rubio H (1989) J Liquid Chromatog 12: 997
24. Brüssau RJ, Stein DJ (1970) Angew Makromol Chem 12: 59
25. García-Rubio LH, Hamielec AM, Gregor JF (1982) In: Computer applications in applied polymer science, ACS Symposium Series, vol 197, Washington, p 151
26. Stützel B, Miyamoto T, Cantow HJ (1972) In: Book of abstracts, Intern Symposium on Macromol. Chem Helsinki 1972, IUPAC, p 337
27. Lange H, Baumann H (1970) Angew Makromol Chem 14: 25
28. Mendelson RA (1986) Proceedings ACS Div Polym Materials 54: 180
29. Samay G, Kubin M, Podesva J (1978) Angew Makromol Chem 72: 185
30. Glöckner G, Stickler M, Wunderlich W (1987) Fresenius Z Anal Chem 328: 76
31. Wunderlich W (1972) J Polym Sci C Polym Symp 39: 145
32. Provder T, Rosen EM (1970) Separation Sci 5: 437
33. Meunier JC, Gallot Z (1972) Makromol Chem 156: 117
34. Park WS, Graessley WW (1977) J Polym Sci A2, Polym Phys 15: 71
35. Kolinsky M, Janča J (1974) J Polym Sci A1, Polym Chem 12: 1181
36. Ambler MR (1980) J Appl Polym Sci. 25: 901
37. Cowie JMG, Maconnachie A (1973) Polym Enging & Sci, vol 13 ref 53, p 401
38. Kamide K, Terakawa T, Miyazaki Y (1979) Polym J 11: 285
39. Rudin A, Hoegy HLW (1972) J Polym Sci A1, Polym Chem 10: 217
40. Valtassari L, Saarela K (1975) Paperi Ja Puu 57: 5
41. Wigand G (1989) personal comm
42. Samuels RJ (1977) Polymer 18: 452
43. Ilchmann D (1982) Thesis, Dresden Univ Technol p 104
44. Echari J, Iruin JJ, Guzmán GM, Ansorena J (1979) Makromol Chem. 180: 2749
45. Iruin JJ, Guzmán GM, Ansorena J (1981) Makromol Chem 182: 2789
46. Morris MC (1970) 9th Intern GPC Seminar 603
47. Gerrens H, Ohlinger H, Fricker R (1965) Makromol Chem 87: 209
48. Kraus G, Stacy CJ (1972) J Polym Sci A2, Polym Phys 10: 657
49. Podesva J, Kratochvíl P, Bohdanecky M, Samay G (1977) J Polym Sci A2, Polym Phys 15: 1521
50. Teramachi S, Hasegawa A, Akatsuka M, Yamashita A, Takemoto N (1978) Macromolecules 11: 1206
51. Teramachi S, Hasegawa A, Takemoto N (1985) Macromolecules 18: 347
52. Elsdon WL, Goldwasser JM, Rudin A (1982) J Polym Sci A1, Polym Chem 20: 3271
53. Wigand G (1989) 5th Seminar on Polymer Physics, KMU Leipzig, Schloß Eyba
54. Kotaka T, Tanaka T, Ohnuma H, Murakami Y, Inagaki H (1970) Polym J 1: 245
55. Minatono S (1966) Thesis, Kyoto Univ. Kyoto, Japan
56. Janča J, Kolinsky M (1977) J Appl Polym Sci 21: 83
57. Janča J, Pokorny S, Kolínsky M (1979) J Appl Polym Sci. 23: 1811
58. Kalal J, Marousek V, Svec F (1974) Angew Makromol Chem. 38: 45
59. Chen HR, Blanchard LP (1972) J Appl Polym Sci 16: 603
60. Kuzaev A, Safonov G, Kudin T (1980) Vysokomol Soedin, Ser A 22: 2088
61. Xu ZD, Song MS, Hadjichristidis N, Fetters LJ (1981) Macromolecules 14: 1591
62. Anderson JN, Barzan ML, Adams HE (1972) Rubber Chem and Technol 45: 1270
63. Moore WR, Uddin M (1969) Europ Polym J 5: 185
64. Dobrowski Z (1984) J Appl Polym Sci 29: 2683
65. Schulz GV, Horbach A (1959) Makromol Chem 29: 93
66. Subramaniam A (1971) Preprints ACS Div Rubber Chem. October, paper 78
67. Mahabadi K, O'Driscoll KF (1977) J Polym Sci 21: 1283
68. Szesztay M, Tüdös F (1981) Polym Bull 5: 429
69. Provder T, Woodbrey JC, Clark JH (1971) Separation Sci 6: 101
70. Chee KK (1985) J Appl Polym Sci 30: 1323
71. Dobbin CJB, Rudin A, Tchir FM (1980) J Appl Polym Sci 25: 2985
72. Mahabadi K Kh (1984) J Polym Sci A2, Polym Phys 22: 449
73. Katime J, Cesteros LC, Ochoa JR (1982) Polym Bull 6: 447
74. Lapin SB, Artemichev VM, Kolegov VJ, Potapov VN, Samarin AF (1987) Vysokomol Soedin Ser A 29: 188
75. Steiskal J, Janča J, Kratochvíl P (1976) Polym J 8: 546

76. Fee JG, Port WS, Whitnauer LP (1958) J Polym Sci 33: 95
77. Scholtan W, Lie SY (1967) Makromol Chem 108: 104
78. Mays JW, Ferry WM, Hadjichristidis N, Funk WG, Fetters LJ (1986) Polymer 27: 129
79. Cane F, Capaccioli T (1978) Europ Polym J 14: 185
80. Ciferri AC, Kryszewski M, Weil G (1958) J Polym Sci 27: 167
81. Sitaramaiah G, Jacobs D (1970) Polymer 11: 165
82. Takahashi A, Ohara M, Kagawa I (1963) Kogyu Kagaku Zasshi 66: 960
83. Freeman M, Manning PP (1964) J Polym Sci A2, Polym Phys 2: 2017
84. Lyngaae-Jorgensen J Chromatog Sci 9: 331
85. Bohdanecky M, Solc K, Kratochvíl P, Kolínsky M, Pyska M, Lim D (1967) J Polym Sci A2, Polym Phys 5: 343
86. Nesterov V, Krasikov VD, Chubarova EV, Turkova LD, Gankina ES, Belenkii BG (1978) Vysokomol Soedin, Ser A 20: 2320
87. Reddy CR, Kashyap AK, Kalpagam V (1977) Polymer 18: 32
88. El'Chibaeva ZS, Bakanova Z Kh, Dzhumadilov TK, Bekturov EA (1982) Izv Akad Nauk Kaz SSR, Ser Khim 6: 51
89. Scholtan W, Marzolph H (1962) Makromol Chem 57: 52
90. Shimura Y (1967) Bull Chem Soc, Japan 40: 273
91. Reddy CR, Kalpagam V (1976) J Polym Sci A-2, Polym Phys 14: 749
92. Frind H (1954) Faserforsch Textiltechn 5: 540
93. Cleland RL, Stockmayer WH (1955) J Polym Sci 17: 473
94. Bisschops J (1955) J Polym Sci 17: 81
95. Misra GS, Mukherjee PK (1978) Colloid Polym Sci 256: 1027
96. Inagaki H, Hayashi K, Matsuo T (1965) Makromol Chem 84: 80
97. Kobayashi H (1959) J Polym Sci 39: 369
98. Ritscher TA, Elias HG (1959) Makromol Chem 30: 48
99. Overberger CG, Pearce EM, Mayes N (1959) J Polym Sci 34: 109
100. Sato H, Yamamoto T (1959) Nippon Kagaku Zasshi 80: 1393
101. Arichi S (1965) Bull Chem Soc, Japan, according to Kurata M et al. in Polymer Handbook, Brandrup J, Immergut EH (eds) Interscience Publishers, New York, p IV–19
102. Bekturov EA, Kamzhamulina RE (1976) Izv Akad Nauk Kaz SSR Ser Chim 26: 30

4 Special Features of Polymer HPLC

In liquid chromatography, the behaviour of synthetic polymer molecules deviates in several respect from the behaviour of low molecular-weight compounds. The differences are caused by some peculiarities of polymers: (i) the small diffusion coefficients of macromolecules in solution, (ii) the size of the macromolecules, which is often of the same order of magnitude as the pore diameter, (iii) the retention of polymers via "trains" of numerous repeat units, (iv) the flexibility of chain molecules, which enables conformational changes to occur, and (v) the limited solubility of polymers. These differences are responsible for the unique behaviour of polymers in gradient elution.

4.1 Diffusion Coefficients of Macromolecules in Solution

The translational diffusion coefficients D_0 of macromolecules are of the order 10^{-7} cm^2/s at MW $= 10^6$. A compilation has been given by Klärner and Ende [1]. The diffusion coefficient increases with decreasing MW. For PS in THF at 30 °C, the relation based upon results from dynamic light scattering reads [2]

$$D_0 = 3 \times 10^{-4} M^{-0.549} \tag{4.1}$$

This relation is shown in Fig. 4.1, line "a". Measurements with PMMA in acetone at 20 °C [3, 4] yielded line "b" with an exponent of -0.58 and a pre-exponential factor 8×10^{-4}. Similar relations are given for further systems in the compilation mentioned [1].

The diffusion coefficients of low MW solutes are of the order 10^{-5} cm^2/s. The extrapolation towards $M = 100$ of the straight lines shown in Fig. 4.1 approximately meets this range.

The symbol D_0 indicates extrapolation to zero concentration. The concentration dependence is given by

$$D = D_0(1 + k_D c) \tag{4.2}$$

Data of k_D for PS in THF are listed in Table 4.1 [5].

The slow diffusion of macromolecules complicates chromatographic separation. Thus, the use of packing materials of small particle diameter d_p is indispensable in polymer chromatography. These materials offer the advantage of shorter diffusion distances and are in general a countermeasure against kinetic peak broadening. Snyder and Stadalius [6] have estimated the optimum particle

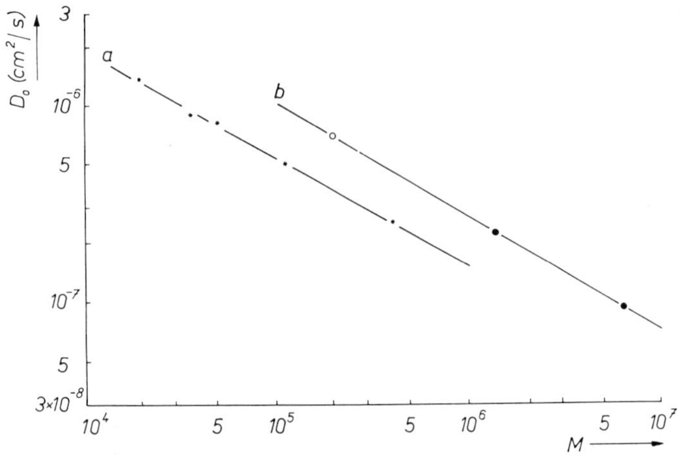

Fig. 4.1. Molecular-weight dependence of diffusion coefficient at zero concentration. (*a*): Polystyrene in THF, 30 °C, *straight line* according to Eq. (4.1), [2]; (*b*): polymethyl methacrylate in acetone, 20 °C, *dots* and *circle* from Refs. [3] and [4], respectively

Table 4.1 Concentration dependence of diffusion coefficients (values of k_D in Eq. 4.2) for polystyrene in tetrahydrofuran at 30 °C [5]

M_w	k_D (cm³/g)
180,000	36 ± 8
390,000	96 ± 7
3,000,000	390 ± 9
10,000,000	620 ± 30

diameter for polymer separations and found values as low as, e.g. $d_P = 0.4\ \mu m$ for denaturated protein of $M = 100,000$. These are calculated values. Experiments under such conditions are hampered by (i) the high flow resistance of packings of this kind, (ii) the fact that polymer samples are likely to block these columns, and (iii) the requirement that the diameter of pressure-resistant particles must exceed the pore diameter by a factor of 10 at least, i.e. packing materials with $d_0 \geqq 300$ nm require $d_P \geqq 3\ \mu m$.

4.2 Stokes Radii and Hydrodynamic Dimensions of Macromolecules

The end-to-end distance R of a carbon chain can be calculated from the bond length (0.154 nm) and the bond angle (109.5°):

$$\langle R^2 \rangle_f^{0.5} = 0.218 n^{0.5} \tag{4.3}$$

where n is the number of carbon atoms in a chain. (Pointed brackets indicate averaging and the index f free rotation around each C—C bond.) The end-to-end distance is given as the root of the averaged squares of all possible conformations and measured in nm. For polymers with two carbon atoms per repeat unit in the main chain, e.g. vinyl polymers, n is related to the degree of polymerization, P, by $n = P*2$. Thus,

$$\langle R^2 \rangle_f^{0.5} = 0.308 P^{0.5} \tag{4.4}$$

or, for polystyrene with molecular weight M,

$$\langle R^2 \rangle_f^{0.5} = 0.0302 M^{0.5} \tag{4.5}$$

With real polymers, rotation is hindred by atoms fixed to the carbon backbone, and sidegroups. The hindrance parameter σ indicates the enlargement of a coil molecule due to restricted rotation. The data in column 3 of Table 4.2 have been calculated by using $\sigma = 2.3$, which is valid for polystyrene. These data represent the unperturbed dimensions of coils (indicated by subscript zero) which can be measured directly in *θ-systems*. Here, the end-to-end distance R is related to the radius of gyration, S, by $\langle R^2 \rangle_0 = 6 \langle S^2 \rangle_0$. These data are listed in column 4 of Table 4.2. In good solvents, polymer coils are even more expanded due to interactions with the solvent. (In reality, polymer coils are ellipsoids with three axes of different length. Thus, the dimensions discussed here are the radii of equivalent spheres.)

According to Stokes, the friction factor f of spheres with radius r_S in a fluid with viscosity η is

$$f = 6\pi\eta r_S \tag{4.6}$$

The friction factor is related to the diffusion coefficient D by

$$f = \frac{RT}{DN_L} = \frac{kT}{D} \tag{4.7}$$

Table 4.2 Molecular size of polystyrene in tetrahydrofuran at 30 °C, calculated through Eqs. (4.5), (4.9), and (4.10)

$10^{-3} M$	$\langle R^2 \rangle_f^{0.5}$ (nm)	$\langle R^2 \rangle_o^{0.5}$ (nm)	$\langle S^2 \rangle_o^{0.5}$ (nm)	r_s (nm)	$[\eta] M\, 10^{-6}$ (ml/mol)	r_h (nm)
10	3.02	6.94	2.83	2.65	0.089	2.42
50	6.75	15.53	6.34	6.42	1.40	6.06
100	9.55	21.96	8.96	9.39	4.59	9.00
200	13.50	31.05	12.68	13.74	15.05	13.36
400	19.09	43.92	17.93	20.11	49.35	19.85
600	23.39	53.79	21.96	25.12	98.83	25.02
800	27.00	62.11	25.35	29.42	161.8	29.49
1000	30.19	69.44	28.35	33.26	237.1	33.50
2000	42.70	98.20	40.09	48.65	777.3	49.77

where R is the gas constant, k the Boltzmann constant, N_L Avogadro's number, and T the absolute temperature. Thus, the **Stokes radius,** r_S, can be calculated by combination of Eqs. (4.6) and (4.7)

$$r_S = \frac{kT}{6\pi\eta D} \tag{4.8}$$

or for THF at 30 °C ($\eta = 0.438$ mPas) and r_S measured in nm:

$$r_S = 5.07 \times 10^{-6}/D \tag{4.9}$$

Stokes radii for PS in THF are listed in column 5 of Table 4.2.

According to Einstein's viscosity law, the **hydrodynamic radius** r_h can be calculated from the hydrodynamic volume V_h which is easily available and widely used for universal calibration in SEC, see Eq. (3.1). Data of V_h, calculated with the help of the Kuhn–Mark–Houwink equation (3.3) and $K_\eta = 12.5\,\mu l/g$, $a = 0.713$ (Ref. [34] in Table 3.1) are compiled in column 6 of Table 4.2. The large values are due to the high MW of the samples and the fact that the coil density in polymer solutions is rather low (order of magnitude about 0.01 g/cm^3).

From the *hydrodynamic volume* (see Sect. 3.1), the hydrodynamic radius of a particle can be estimated by a straightforward geometric approach

$$r_h = \left(\frac{3 \times 10^{21}[\eta]M}{10\pi N_L}\right)^{1/3} = 0.054([\eta]M)^{1/3} \tag{4.10}$$

which easily can be applied also to copolymers. Equation (4.10) yields an average radius, corresponding to the average values of M and $[\eta]$. In Table 4.2, column 7 lists r_h calculated by Eq. (4.10). Almost the same values result from a more fundamental approach based upon the Flory–Fox and the Ptitzyn–Eisner equations [7].

From columns 4, 5, and 7 it can be concluded that, in the samples to be investigated, the size of macromolecules is in the range of 5–100 nm. If unrestricted permeability is required, the pore diameter d_0 of a chromatographic packing should exceed the Stokes diameter d_S of the solute by more than a factor of three. Size exclusion occurs if $d_S \leqq d_0 \leqq 3d_S$. In order to avoid superimposition of SEC phenomena and gradient HPLC, the latter should be performed on packings with either $d_0 \geqq 300$ nm or $d_0 \leqq 5$ nm.

Experimental proof of this statement has been published by Mori et al. [8] who showed that on silica columns S/MMA copolymers ($10^{-3}M_w = 120 - 150$) eluted in broad and multimodal peaks if $d_0 = 10$ nm, but in sharp peaks if $d_0 = 3$ nm, see Fig. 4.2.

The elution characteristic, i.e., the eluent composition at peak maximum vs copolymer composition, is only slightly influenced by the question of whether or not a polymer sample can penetrate the packing material. This has been found with S/MMA copolymers on silica with pore size graded by two orders of magnitude (exclusion limits $3 - 1000 \times 10^4$) [9].

The use of porous particles permeable to solute molecules is common praxis in HPLC. Several difficulties encountered with polymers due to their large size

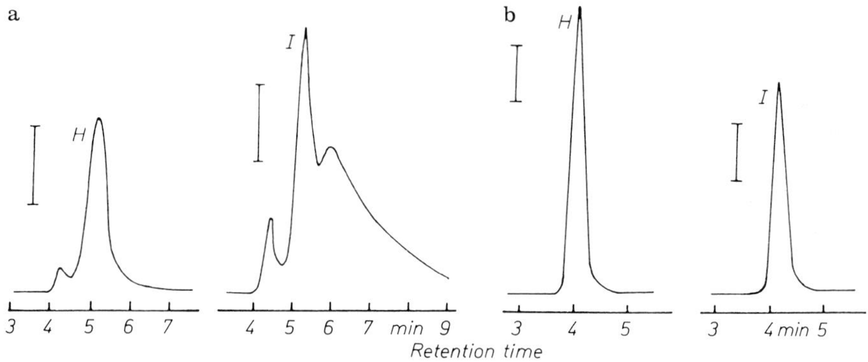

Fig. 4.2 a, b. Effect of pore diameter on the elution of *stat*-copoly (styrene/methyl methacrylate) samples. Column 250 × 4.6 mm, silica packing, $d_P = 5\,\mu m$. (**a**) $d_o = 10$ nm, $m_o = 66\,\mu g$, $V_o = 33\,\mu l$; (**b**) $d_0 = 3$ nm, $m_0 = 34\,\mu g$, $V_0 = 17\,\mu l$; eluent trichloromethane (with about 1% ethanol), flow rate 0.5 ml/min; copolymers; H (33.7 mol% MMA), I (42.6%), see Table 9.3, $M_w \times 10^{-3} = 120–150$, $M_n \times 10^{-3} = 70–100$. Bars indicating 64 mAU. From Ref. [8] with permission

and low diffusion coefficients has brought up the idea of performing polymer HPLC on column packings impermeable to large molecules. In fact, small-pore packings ($d_0 \leqq 5$ nm) proved efficient [10, 11]. Unger et al. successfully introduced even nonporous spheres for polymer separations [12–14].

4.3 Retention by Multiple Attachment

In interaction chromatography of polymers with *isocratic elution* the strength of the eluent has a dramatic effect on retention. With a weak eluent, the retention time of a polymer usually exceeds by far any reasonable period of experimental work. Increase of the elution strength will eventually lead to a sudden change to the opposite behaviour: the polymer is not retained at all and leaves the column at *dead time t'*. A minute alteration of elution conditions causes transition from zero retetion to infinity. This "on or off" behaviour produces the impression of irreversible fixation under the conditions of retention.

This paragraph shows that the strange behaviour can be understood as a consequence of multiple attachment [15]. Synthetic polymers consist of a large number of repeat units. In principle, all of them have the chance of becoming adsorbed but the conformation of adsorbed polymer coils usually comprises loops and tails extending in the solution and trains of adsorbed repeat units.

The *capacity factor k'* is a measure of the probability ratio for a given solute to be either retained or not. With the probability p of being retained, the probability of being movable is $(1-p)$ and

$$k' = p/(1-p) \qquad (4.11)$$

or

$$1 - p = \frac{1}{1+k'} \qquad (4.12)$$

In *isocratic elution*, the optimum value of the capacity factor is $k' = 2$. In this case, the solute remains in the stationary phase twice as long as in the mobile phase. The likelihood of mobility is $1/(1 + k') = 0.33$ which, under identical chromatographic conditions, should also hold for an isolated repeat unit of the same structure.

A polymer chain is retained in the stationary phase as long as one of its repeat units is adsorbed. A chain can migrate only if all constituting units are in the mobile phase. Assuming independent adsorption/desorption equilibrium for each unit, one can express the mobility condition of a macromolecule by the product of the corresponding probabilities of repeat units, $(1 - p_u)$. This yields for a train of P units

$$(1 - p_{total}) = (1 - p_u)^P \tag{4.13}$$

and the capacity factor becomes:

$$k'_{total} = (k'_u + 1)^P - 1 \tag{4.14}$$

With real polymers, P is rather large.

If $k'_u = 0$, then k'_{total} is zero as well, i.e. if the retention of an isolated unit is zero the whole polymer, irrespective of its size, is also eluted without any retention. A minute decrease of elution strength may cause some retention of a repeat unit, e.g. $k'_u = 0.05$. This means that the elution time of a monomer with the structure of a repeat unit would increase by 5%, e.g. from 2 min to 126 s. If we assume $P = 100$, Eq. (4.14) yields $k'_{total} = 130.5$. Thus, the same change in elution strength which is just noticable in low-molecular HPLC causes a polymer to be retained on the column for more than four hours. (Experimental evidence of behaviour of this kind is reported in Sect. 6.1.)

Elution conditions which are normal for low-molecular solutes, e.g. yielding $k' = 2$, would cause a polymer with an adsorbed train of 100 units to stick in the column for centuries (by Eq. (4.14), $k'_{total} = 5 \times 10^{47}$).

4.4 Conformation of Synthetic Polymers

The macromolecules of soluble synthetic polymers have the structure of chains (possibly with branches). Special effects aside, the most probable conformation is a random coil structure which, in dilute solutions, embraces about 99% solvent. The flexibility of the chains easily enables conformational changes in solution. Under the conditions of high-pressure liquid chromatography, random coils may be deformed in a way that long trains of units come in close contact with the surface of the stationary phase and only a few loops and tails protract into the mobile phase. In pores, coiled macromolecules may adjust their conformation to the size of the cavity in order to get the maximum of interaction enthalpy.

The flexibility can cause dramatic differences in chromatographic behaviour by comparatively small changes in eluting conditions. For instance, Knox and McLennan [16] found a steep increase in plate height at very low flow rate for PS ($M = 200{,}000$) in DCM on columns packed with silica, $d_0 = 12$ nm. At a superficial glance, it could have appeared like the behaviour predicted by the *van Deemter equation*, but the minimum occurred at far too high values of reduced velocity. The position of the minimum changed towards higher reduced velocity with increasing column length and molecular weight of the samples. With the increase of plate height at very low velocities, the peaks became tailed. Thus it was concluded that the surprising effect was caused by a slow unfolding process of polymer coils, possibly to be interpreted as partial creeping of the flexible macromolecules into pores which are too small for penetration [16].

Polystyrene of $M = 173{,}000$ and coil diameter 30 nm has been reported to be capable of entering pores of silica gel with 10 nm average pore diameter [17].

4.5 Solubility of Polymers

Solubility generally requires a negative change in Gibbs' free energy function G on mixing, i.e.

$$\Delta H_{\mathrm{m}} - T\Delta S_{\mathrm{m}} = \Delta G_{\mathrm{m}} < 0 \tag{4.15}$$

(ΔH_{m}: Heat of mixing; ΔS_{m}: Entropy of mixing)

The dissolution of low-molecular compounds yields a substantial increase in entropy: the transition from the highly ordered three-dimensional structure of a crystal to the random distribution in a solution is accompanied by a considerable change in entropy. In low-molecular mixing processes, the contribution $-T\Delta S_{\mathrm{m}}$ is so large that even positive values of ΔH_{m} usually do not prevent dissolution. With macromolecules the situation is quite different because polymers normally have low order in solide state. In addition to this, the regular arrangement of repeat units along a chain remains on dissolution. Considering the solvent, one will possibly experience even an increase in order, which is due to solvatization of macromolecules. Furthermore, in solutions of equal weight concentrations, the number of solute particles is much greater in low-molecular than in macromolecular systems. Thus, an entropy contribution to ΔG_{m} will be small and the condition $\Delta G_{\mathrm{m}} < 0$ requires a negative change in enthalpy, $\Delta H_{\mathrm{m}} < 0$.

In terms of Hildebrand parameters, the δ value of a solvent must be very close to that of the polymer of interest. Thus, the list of solvents for a given polymer is rather short and most of the easily available liquids are nonsolvents, see the compilation given by Fuchs and Suhr [18]. In contrast to this, low MW solutes usually have a broad variety of potential solvents according to the span of acceptable δ values.

The Flory–Huggins theory of polymer solutions gives a quantitative description of the thermodynamic properties of these systems. From the original formulation of this theory, the activity of the homogeneous polymer solute is

$$\ln a_2 = \ln \Phi_2 + (1 - V_2/V_1)\Phi_1 + (V_2/V_1)\chi\Phi_1^2 \tag{4.16}$$

where V_1, V_2 are the molar volume of solvent and solute, Φ_1, Φ_2 the volume fraction of solvent and polymeric solute, respectively, and χ the Flory–Huggins interaction parameter which accounts for a free energy contribution due to nearest-neighbour interactions.

Thus, by setting $\ln a_2 = 0$ and $\Phi_1 \to 1$, the volume fraction of a polymer in solution can be calculated which is in equilibrium with a slightly swollen polymer:

$$\ln \Phi_2 = (V_2/V_1) - 1 - (V_2/V_1)\chi \tag{4.17}$$

or, taking $-V_2/V_1$ as a factor,

$$\ln \Phi_2 = -(V_2/V_1)[\chi + (V_1/V_2) - 1] \tag{4.18}$$

The molar-volume ratio of polymer to solvent, V_2/V_1, increases linearly with the degree of polymerization, P. The inverse ratio, V_1/V_2, is very small. Thus, the sum given in square brackets in Eq. (4.18) is almost constant and the solubility of polymers decreases with chain length approximately by

$$\Phi_2 = \exp(-\text{const}\, P) \tag{4.19}$$

With copolymers, the value of χ is, at intermediate composition, smaller than χ of the respective parent homopolymers; this way, the increase in solubility of statistical copolymers can be understood.

4.6 Gradient Elution of Polymers

Polymer samples which are homogeneous in chemical nature and chain length can be eluted isocratically. Samples which consist of species differing substantially in degree of polymerization or chemical structure cannot be separated isocratically because the different species would exhibit capacity factors which vary too much. The immense dependence of polymer retention on elution strength discussed in Sect. 4.3 requires *gradient elution*. For reviews of gradient HPLC, see [6, 19, 20].

The *solute acceleration factor* S which is between 1 and 5 for low-molecular solutes, can be 30 to 100 times larger with polymers [21]. The high S values of polymers can be understood on base of a multiple-attachment model described in Sect. 4.3. From Eq. (12.44), it follows for a repeat unit

$$k'_{u,A}/k'_u = 10^{S_u \Phi_B} \tag{4.20}$$

(S_u: solute acceleration factor per repeat unit)
Combining Eqs. (4.14) and (4.20) and considering $(k'_u + 1)^P \gg 1$, one obtains the approximation

$$S_{\text{total}} \approx P \cdot S_u \tag{4.21}$$

i.e., the S values of polymers are larger than those of low-molecular solutes and should increase with degree of polymerization. An increase of S with MW has been established experimentally with PS oligomers [22, 23]. From Table IV in Ref. [22], it follows for oligomers: $S_{\text{total}} = 3.95 + 0.38P$. For a wider range of molecular weight, an increase according to $S_{\text{polymer}} = 0.22M^{0.5}$ has been reported [24].

4.7 Gradient Elution of Polymers on Small Pore Packings

This model presupposes that the polymer solute is completely excluded from the pores of a packing. Only the interstitial volume V_I is accessible for the polymer. When in the mobile phase, the polymer solute is transported at a linear velocity

$$u_P = L/(V_I/F) \qquad (4.22)$$

where L is the length of the column and F the flow rate. In contrast to the polymer, the eluent has access to the pore volume of the packing. The eluent volume in the column is V_{mob}, which is the sum of V_I and the total pore volume of the packing. This gives the linear velocity of the eluent:

$$u_E = L/(V_{mob}/F) \qquad (4.23)$$

Since $V_{mob} > V_I$, the velocity of the eluent is smaller than the velocity of the polymer. The polymer bypasses the pores and thus always overtakes the eluent having sufficient elution strength. The polymer is retained again until a more powerful eluent reaches its position.

This model gives the opportunity for multi-stage separation which is fundamental in (isocratic) LC. The model was first described [25] with respect to a precipitation/re-dissolution mechanism but, in principle, it does not include a certain mechanism. If the polymer precipitates by rushing into the poor solvent ahead, the segregated gel droplets must stick to the stationary material in the column and must not remain in the interstitial volume, see Sect. 8.4.

4.8 The Martin Equation

For the members of a homologous series, a constant contribution to retention on increasing the chain length by a carbon atom was found by Nobel laureat A.J.P. Martin in 1949 [26]. This can, with carbon number n and group-specific contribution C_1, be written as

$$\log k'_{Martin} = C_1 + C_2 n \qquad (4.24)$$

Figure 4.3 shows that polystyrene adheres to the Martin equation for $P \leqq 19$ (equivalent to n = 38 carbon atoms in the main chain). At higher P, the k' data bend off the straight-line plot obtained from low P values. This can be understood by the effect of tails and loops which protrude in the mobile phase and diminish the attachment of the trains, especially through mechanical forces exhibited by the streaming liquid. (Figure 4.3b contains data up to $\log k' = 13.4$ for $P = 480$ or $M = 50{,}000$ which have been calculated from gradient elutions.)

The data used for Fig. 4.3 were measured with W/THF mixtures. Adherence to the Martin equation has also been proved for PS oligomers in Hex/THF and W/AcN eluents [23].

Note the similarity between the Martin equation and the logarithmic forms of Eqs. (4.14) or (4.19). The latter has been derived from solubility equilibrium, the former from adsorption/desorption effects. Thus, adherence to the Martin equation is not automatically the proof of a specific mechanism of retention.

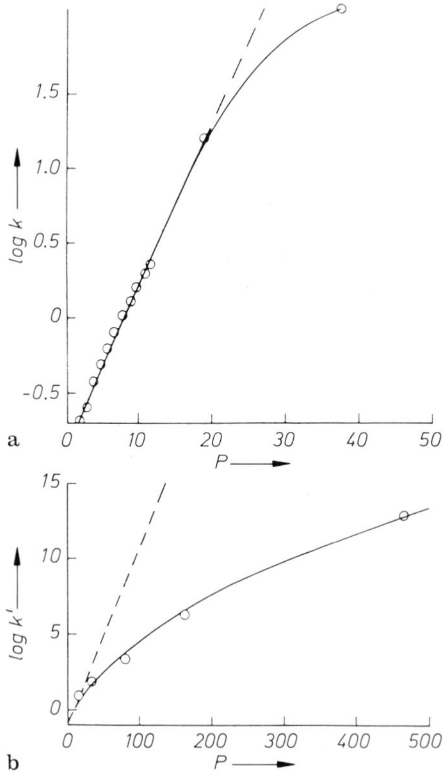

Fig. 4.3 a, b. Verification of the Martin equation for styrene oligomers. Column (250 × 4.6mm) packed with RP C18 based on pellicular silica ($d_0 = 100$ nm, $d_p = 30$ μm), eluent: water/THF (40:60); degree of polymerization (**a**): 2–40 (isocratic elution), (**b**): 19–470 (data estimated from gradient elutions). From Ref. [22] with permission

4.9 Effect of Block Structure on Retention

Experimental results published so far indicate longer retention of *block copolymers* than of *statistical copolymers* (same composition and similar MW provided). The systems investigated are S/Bd [27], S/TBMA [28], and DMA/MMA [29]. The coincidence of the results can be understood on the basis of a model discussed in Sect. 4.3, which also yielded Eq. (4.20). The latter enables the volume fraction of a strong eluent to be calculated which is needed for the elution of a polymer sample after multi-site retention. With reasonable estimates for the *solute acceleration factor S*, the *capacity factor* $k'_{u,A}$ for the retention of a monomeric unit in the starting eluent, and the capacity factor k'_{total} at the elution of a multiply retained polymer, Eq. (4.20) yields, in combination with Eq. (4.14), the straightforward result that elution requires the higher a volume fraction of strong eluent the longer the adsorbed trains are.

Chromatographic separation according to composition of copolymers requires different retention of constitutionally differing units. Under extreme (but reasonable) HPLC conditions, only one kind of structural units (e.g. "A") is retained at all and the others not. Thus, the exponential relation between the number of neighbouring units and the capacity factor requires sequences with a corresponding number of A-units in a copolymer. Uninterrupted sequences of that kind can be found in block copolymers but (apart from exotic systems with strange values of *monomer reactivity ratios*) certainly not in statistical copolymers where microblocks of only a few repeat units prevail, i.e. rather short sequences of A-units separated from each other by sequences of the complementary units which are not retained under the conditions of the experiment. Hence, augmented retention through multiple attachment is more feasible with block copolymers than with statistical ones. Indeed, experimental results indicate stronger retention of block copolymers in comparison with statistical ones of the same chemical composition.

4.10 References

1. Klärner PEO, Ende HA (1975) In: Brandrup J and Immergut EH (eds) Polymer Handbook, 2nd edn, John Wiley, New York, p IV/61
2. McDonnell ME, Jamieson AM (1977) J Macromol Sci-Phys B12: 67
3. Lütje H, Meyerhoff G (1963) Makromol Chem 68: 180
4. Meyerhoff G (1967) J Polym Sci C16: 1579
5. Yu TL, Reihanian H, Jamieson AM (1980) Macromolecules 13: 1590
6. Snyder LR, Stadalius MA (1986) In: Horvath Cs (ed) High-performance liquid chromatography, 4: 195
7. Eltekov YA (1986) J Chromatogr 365: 191
8. Mori S, Uno Y, Suzuki M (1986) Anal Chem 58: 303
9. Danielewicz M, Kubin M (1981) J Appl Polym Sci 26: 951
10. Termachi S, Hasegawa A, Shima Y, Akatsuka M, Nakajima M (1979) Macromolecules 12: 992
11. Glöckner G, Kroschwitz H, Meissner C (1982) Acta Polymerica 33: 614
12. Unger KK, Jilge G, Kinkel JN, Hearn MTW (1986) J Chromatogr 359: 63
13 Jilge G, Janzen R, Giesche H, Unger KK, Kinkel JN, Hearn MTW (1987) J Chromatogr 397: 71
14. Janzen R, Unger KK, Giesche H, Kinkel JN, Hearn MTW (1987) J Chromatogr 397: 91
15. Glöckner G (1986) In: Advances in Polym Sci, 79: 159
16. Knox JH, McLennan F (1979) J Chromatogr 185: 289
17. Belenkii BG (1979) Pure Appl Chem 51: 1519
18. Fuchs O, Suhr HH (1975) In: Brandrup J and Immergut EH (eds) Polymer Handbook, 2nd edn, John Wiley, New York, p IV/241
19. Snyder LR (1980) In: High-performance liquid chromatography, Horvath C (ed), 1: 207
20. Jandera P, Churácek J (1985) Gradient elution in column liquid chromatography, theory and practice, J Chromatog Library 31, Elsevier, Amsterdam
21. Snyder LR, Stadalius MA, Quarry MA (1983) Anal Chem 55: 1412A
22. Larmann JP, DeStefano JJ, Goldberg AP, Stout RW Snyder LR, Stadalius MA (1983) J Chromatogr 255: 163
23. Lai S-T, Sangermano L, Locke D (1984) J Chromatogr 312: 313
24. Quarry MA, Stadalius MA, Mourey TH, Snyder LR (1986) J Chromatogr 358: 1
25. Glöckner G (1983) Pure Appl Chem 55: 1553
26. Martin AJP (1949) Biochem Soc Symposia (London) 3: 4
27. Sato H, Takeuchi H, Tanaka Y (1985) Int Rubber Conf, Kyoto [18–B15]: 596
28 Glöckner G, Müller AHE (1989) J Applied Polym Sci 38: 1761
29. Müller MA, Augenstein M, Dumont E, Pennewiß H-New Polym Materials (submitted for publ.)

5 Solubility and Adsorption Effects in Polymer HPLC

It has been stated "that normal chromatographic retention and precipitation/redissolution represent two processes, either of which can determine chromatographic retention in a given case" [1]. This chapter deals with the methods of finding out whether or not one of the mechanisms in question predominates. Information on the mechanismus of a given polymer separation can be derived (i) from the dependence of retention time on sample size, (ii) by comparison between solubility and elution characteristics, (iii) from a closer inspection of the chromatographic system that affects separation, and (iv) from the effect of temperature on retention.

5.1 Dependence of Retention Time on Sample Size

When polymer retention is governed by precipitation and redissolution, the characteristics of solubility cause retention time to increase with sample size. The solubility of a given polymer in a solvent/nonsolvent mixture with volume fraction Φ of the solvent and polymer concentration c can be described by the relation

$$\Phi^* = C_1 + C_2 \log c^* \tag{5.1}$$

where a star indicates cloud-point data, i.e. values at the verge of precipitation. Equation (5.1) has been established by turbidimetric titration of, e.g. polycarbonates in DCM/MeOH [2], see Fig. 5.1, and by the measurement of PS concentrations in given mixtures of THF/W [1].

In gradient elution, a higher volume fraction of the stronger eluent B corresponds to later elution. Retention increasing with sample load has been observed with, e.g. S/AN copolymers in *i*Oct/THF gradients on C18 columns [3] and PS samples in MeOH/DCM gradients on silica columns [4], see Fig. 5.2.

The effect of concentration on solubility depends on molecular weight. This can be seen also in Fig. 5.1. Experimentally, an increase of factor C_2 in Eq. (5.1) by approximately $M^{-0.5}$ has been estimated from turbidimetric titrations [2] as well as chromatographic results [5].

In adsorption chromatography, sample size has a negligible effect on retention time as long as adsorption isotherms are linear, but column overload would decrease retention. The latter phenomenon has been observed with, e.g.

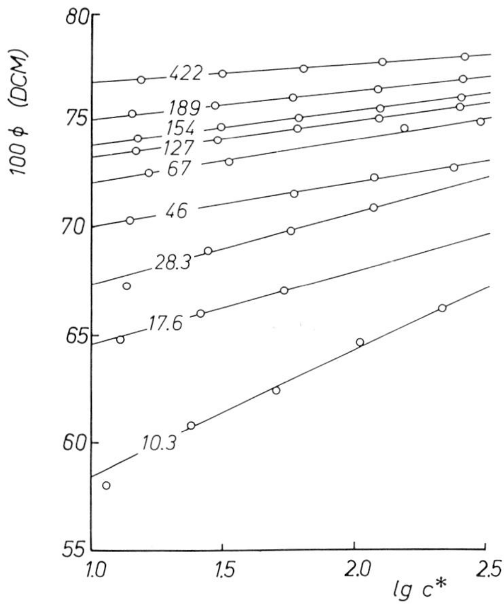

Fig. 5.1. Effect of polymer concentration on cloud point. Turbidimetric titration of polycarbonate in dichloromethane solution by methanol nonsolvent, 20 °C. Concentration in mg/1. (Figures on the *straight lines* indicate sample molecular weight, $10^{-3}M$)

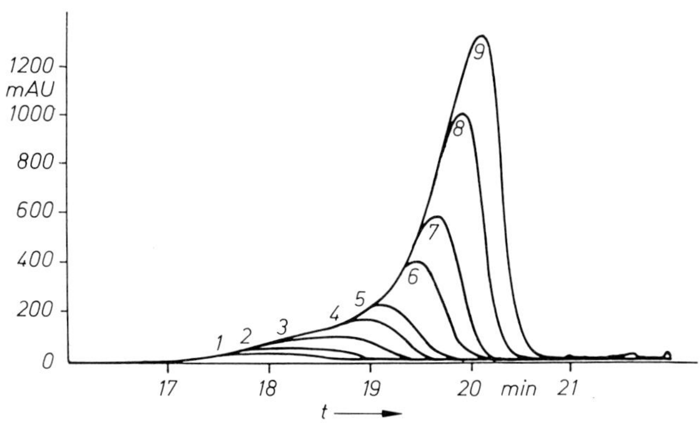

Fig. 5.2. Sample-size effect on the retention of polystyrene in methanol/dichloromethane gradient. Molecular weight 110,000, column RP C18; *1*: $m_0 = 12.5\,\mu g$; *2*: $25\,\mu g$; *3*: $50\,\mu g$: *4*: $75\,\mu g$; *5*: $100\,\mu g$; *6*: $150\,\mu g$; *7*: $200\,\mu g$; *8*: $300\,\mu g$; *9*: $400\,\mu g$. From Ref. [4] with permission

PS in W/THF gradients [1], with PS in Hp/DCM gradients, see Fig. 5.3, and with lactalbumin in isocratic elution [4].

The sample-size criterion is general and can be applied to HPLC of any polymer. An increase of retention with increasing amount injected indicates predominance of a precipitation mechanism, whereas decreasing retention with increasing sample size indicates predominance of an adsorption mechanism.

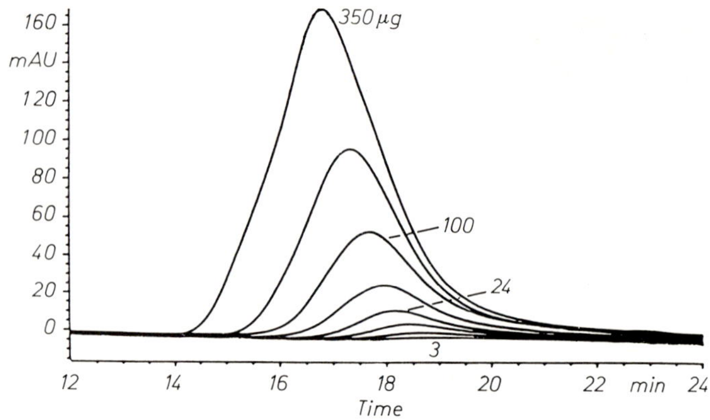

Fig. 5.3. Sample-size effect on the retention of polystyrene in *n*-heptane/dichloromethane gradient. Molecular weight 200,000, column RP C-18, m_0 in μg indicated. From Ref. [4] with permission

5.2 Comparison Between Solubility and Elution Characteristics

When polymer HPLC is governed by adsorption, a sample will elute in an eluent mixture which is a solvent for the sample, i.e. in Fig. 5.1, in the region above the solubility borderline for the respective MW. A comparison between solubility line and elution characteristics yields information not only for an individual sample but for the whole ensemble of polymers whose separation is intended, i.e., for polymer homologues or for different copolymer samples from a given combination of monomers. (Of course, the sample-size criterion could be applied as well but would require the investigation of concentration series for each member of the ensemble. Besides the labour involved, that method is often impracticable, because polymer standards are usually not available in large quantities.)

The direct comparison described in this section saves precious sample material and labour but requires great care with respect to experimental conditions. Due to the logarithmic dependence expressed by Eq. (5.1), the effect of polymer concentration on solubility is especially pronounced at low values of c. For comparison of eluent composition at peak elution with cloud points from turbidimetric titration it is essential that the polymer concentration in both experiments are, at least, of the same order of magnitude.

The width of a Gaussian peak is approximately $6s$ (s: standard deviation). The polymer concentration at the apex of the peak can be approximated by the concentration in a slice between $(V_e - s/2)$ and $(V_e + s/2)$ which contains almost 40% of the substance. For instance, with a peak volume of 0.5 ml on injection of 2 μg, there is 0.8 μg polymer in $0.5/6 = 0.08$ ml eluate, corresponding to a polymer concentration of about 10 mg/l.

Solubility can be measured by titrating a dilute solution of the sample with a nonsolvent which must be miscible with the solvent. The procedure is carried out

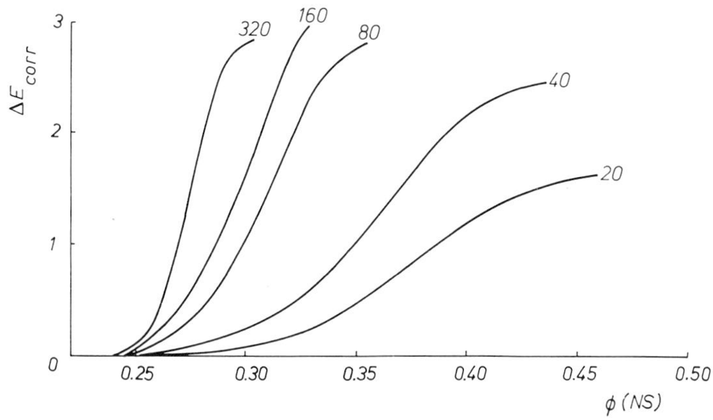

Fig. 5.4. Turbidimetric titration of polycarbonate in dichloromethane. Increase of turbidity on addition of methanol nonsolvent, 20 °C, molecular weight 28,300. Abscissa: volume fraction of methanol, ordinate: apparent extinction, corrected for dilution. Figures at the curves indicate polymer concentration in mg/l

under strict temperature control and monitored by either the scattered or transmitted light. On addition of a certain volume of nonsolvent, the first traces of polymer are precipitated as highly swollen gel. Further addition of nonsolvent causes the turbidity to increase until the polymer is completely precipitated in the gel phase. Although these experiments are rather simple, a strict regime is essential for consistent results since the turbidity is influenced by gel aging and other effects.

The immediate result of **turbidimetric titration** can be given as a plot of turbidity vs volume fraction nonsolvent, see Fig. 5.4. The crude curves must be corrected for dilution. Usually, S-shaped curves are found. The slowly rising initial part of the curves is caused by concomitant polymer of lower solubility, e.g. of higher MW. Narrow fractions yield curves which rise steeply from the baseline, see Fig. 5.5. The intersection of the steepest tangent with the baseline gives a reliable measure of the cloud point of a sample.

The actual concentration of a sample is determined by its initial concentration c_0 and the dilution due to the volume fraction Φ^*_{NS} of nonsolvent added,

$$c^* = c_0/(1 + \Phi^*_{NS}) \tag{5.2}$$

This concentration should equal the sample concentration at peak position, vide supra. Correspondence is also required for temperature and pressure, which affect solubility as well. Experimental obstacles against establishing full correspondence in the latter variables should be compensated by a careful balance of concentration.

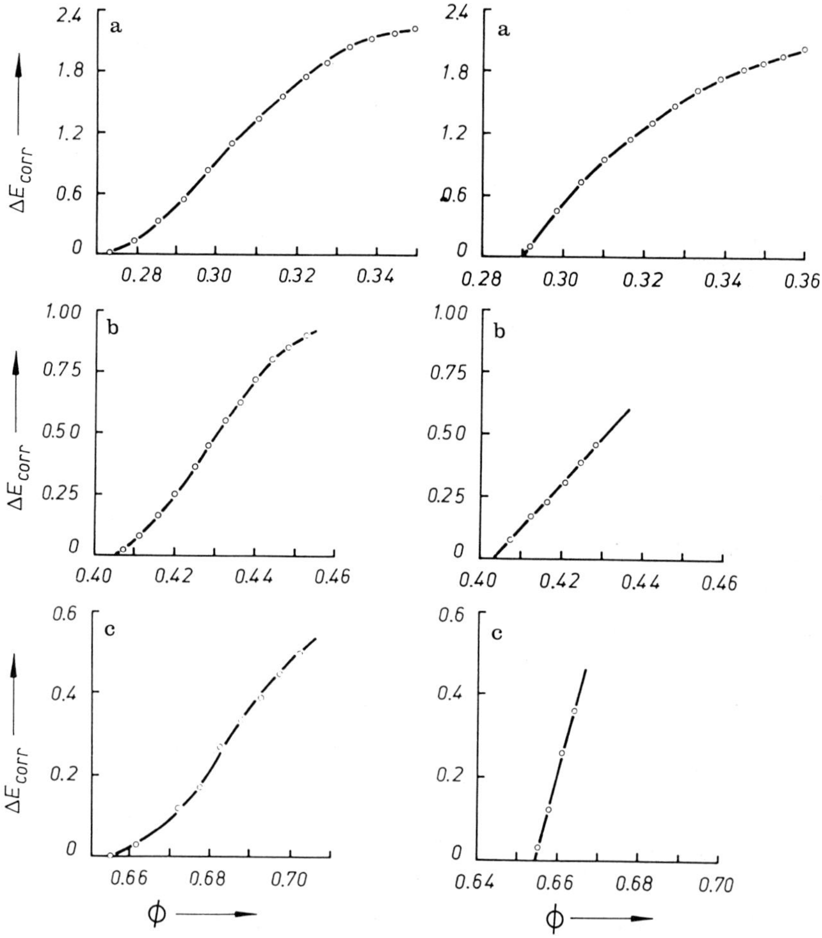

Fig. 5.5 a–c. Turbidimetric titration; influence of sample homogeneity on the steepness of turbidity curves. (ϕ: volume fraction of nonsolvent, NS). Left column of diagrams measured with fractions from precipitation fractionation, right-hand column with fractions from *Baker–Williams fractionation.* (**a**) polycarbonate (MW: 46,000/46,500; data refer to the left/right-hand diagram of each system, respectively) in dichloromethane, NS methanol, (**b**) polycarbonate (MW: 67,000/70,000) in trichloromethane, NS *iso*-propanol, (**c**) poly(butyl methacrylate) (MW: 98,000/98,000) in acetone, NS methanol

5.2.1 Retention According to Molecular Weight

Polymerhomologues are polymers which are identical in composition and structure but unlike in degree of polymerization. Preferably, separation by chain length is performed by SEC which is a versatile and convenient method, but the MW effect in gradient HPLC is also of interest, e.g. as a side effect in separating copolymers by composition. Further to this, it is the central point in HPLC experiments with PS standards, which are commercially available, chemically

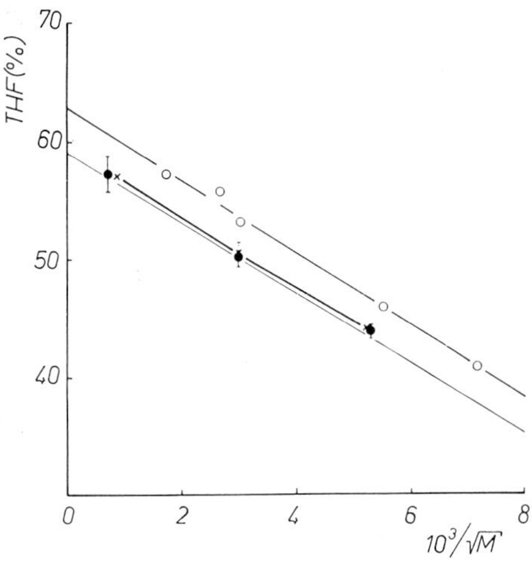

Fig. 5.6. Molecular-weight dependence of solubility and retention of polystyrene standards in methanol/tetrahydrofuran mixtures, plotted according to Eq. (5.3). o: turbidimetric titration of 40 mg/1 solutions in THF with methanol nonsolvent; *light line*: ditto, but polymer concentration 10 mg/l; x, •, and *heavy line*: eluent composition at peak position on C18 column at 50 °C. By courtesy of Friedr. Vieweg & Sohn Verlagsgesellschaft [7]

homogeneous, sufficiently characterized, and rather narrow in chain length distribution.

The volume fraction Φ^*_{NS} of nonsolvent at the cloud point is related to molecular weight M by the empirical equation [6]

$$100\,\Phi^*_{NS} = C_1 + C_2 M^{-0.5} \tag{5.3}$$

which can be understood on the basis of the Flory–Huggins theory. Since $100\,\Phi^* = 100(1 - \Phi^*_{NS})$, a plot of solvent percentage vs $M^{-0.5}$ differs only by the sign of the slope. The advantage of the latter representation is its similarity with Fig. 5.1, i.e. the areas of homogeneous solutions are located above the solubility lines. A molecular-weight effect in polymer HPLC can be also evaluated by plotting the eluent composition vs $M^{-0.5}$, see Sect. 8.5.1.

Figure 5.6 shows the plot of $100\,\Phi_{THF}$ vs $M^{-0.5}$ for PS in MeOH/THF. The straight line determined by open circles represents solubility at $c_0 = 40$ mg/l, the light line at 10 mg/l. The position of the lines indicates a higher content of nonsolvent required for precipitation at a higher sample dilution, in agreement with Eq. (5.1).

The HPLC results precisely meet the solubility characteristics. The crosses and full circles in Fig. 5.6 and their heavy connecting line show the THF concentration at peak elution from a RP C18 column. This data was calculated from elution time and gradient program. Each dot represents at least 10 analyses from which average and standard deviation were calculated. The crosses had been measured with other samples and different brands of eluents two years before. Both sets of HPLC results agree well with the position of the solubility line obtained by TT [7].

A different example is given in Fig. 5.7 which shows corresponding results for the system PS – THF/*i*-Oct. The open circles represent TT results of PS solutions in THF with *i*Oct as a precipitant at a starting concentration $c_0 = 40$ mg/l. The

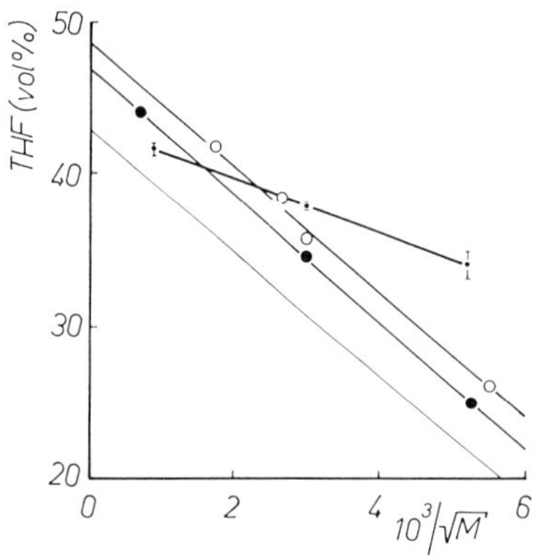

Fig. 5.7. Molecular-weight dependence of solubility and retention of polystyrene standards in *iso*-octane/tetrahydrofuran mixtures, plotted according to Eq. (5.3). ○: turbidimetric titration of 40 mg/l solutions in THF with *iso*-octane nonsolvent, ●: ditto, with addition of 2% methanol; light line: ditto, but polymer concentration 10 mg/l; • and heavy line: elution characteristics on silica column at 50 °C. By courtesy of Friedr Vieweg & Sohn Verlagsgesellschaft [7]

full circles, which determine the somewhat lower line, were also measured at 40 mg/l starting concentration but with *i*Oct + 2% MeOH as a precipitant. This data was needed since the chromatographic experiments were performed with 2% MeOH in the eluent. Analytical-grade *i*Oct as a starting eluent caused ghost peaks due to impurities trapped on the column during the prerun periode. The addition of 2 vol% MeOH proved a remedy against these distortions [8]. The thin line in Fig. 5.7 gives an estimate of what can be expected on reducing the starting concentration of the polymer from 40 to 10 mg/l. (The line is shifted by just the same distance as in Fig. 5.6 where it had been measured.)

The dots and their heavy connection represent HPLC results obtained on a silica column. The deviation between elution characteristics and solubility line indicates that, in this case, retention is determined by adsorption. Note that the slope of the elution characteristics and, hence, the selectivity of separation according to MW is smaller than in the system shown in Fig. 5.6.

It is worthwhile to note that the addition of MeOH nonsolvent, whose precipitating strength is even greater than that of *iso*-octane, increases the solubility of polystyrene: the full circles are below the line determined by the open ones. This shows that less THF is required in combination with *i*Oct–MeOH mixtures than in combination with pure *i*Oct. Similar co-solvency effects are often found with polymers.

5.2.2 Retention According to Copolymer Composition

As a rule of thumb, a solvent/nonsolvent combination will fractionate copolymers by MW if the solvent is capable of dissolving both parent homopolymers (e.g. PS and PMMA for S/MMA copolymers) and the nonsolvent acts in an approximately equivalent manner on both parent polymers. On the other hand, fractionation by composition will take place if a copolymer solvent dissolves only one of the homopolymers. The latter effect can be amplified by the choice of the nonsolvent.

If the constituting units of a copolymer differ substantially in polarity, a nonpolar precipitant will facilitate fractionation by composition. The effect of nonsolvent can even surpass the influence of solvent. For instance, S/AN copolymers in DCM solutions are mainly fractionated according to composition when alkane hydrocarbons are used as precipitants, whereas, in a limited range of copolymer composition ($\Delta w_{AN} \approx 20\%$), MeOH nonsolvent fractionates according to MW [9]. With some qualifications, this fractionation tendency of either hydrocarbons or MeOH also holds for S/AN solutions in DMF, nitrobenzene, acetone, MEK, and dioxane [10].

In correspondence with the effect of hydrocarbons in S/AN precipitation, the HPLC retention of S/AN copolymers increases with AN content in gradient elution by *i*Oct/THF. In this nonsolvent/solvent-system, the elution characteristics plotted as % THF vs w_{AN} is so close to the solubility line measured by TT that retention is obviously due to precipitation.

5.3 Correspondence Between Column Polarity and Gradient Orientation

The comparison of solubility and elution characteristics (discussed in the previous Sects. 5.2.1 and 5.2.2 as well as in 6.2 and 6.3) gives a clear decision only if the elution characteristics are located in the area of homogenous solutions, i.e. above the solubility borderline. This position indicates retention by adsorption. Elution characteristics coincident with the solubility borderline of the system under consideration may indicate a precipitation/redissolution mechanism, but the decision is strongly dependent on the experimental conditions in TT and HPLC measurements, see Sect. 5.2. A rigorous proof can be derived from HPLC experiments with different combinations of mobile and stationary phases.

In 1982, HPLC separations of S/AN copolymers through Hex/THF gradients have been reported, where columns packed with either bare silica or C8 bonded phase material yielded essential the same chromatograms [11]. This surprising observation could be confirmed in subsequent studies [12, 13]: S/AN samples of higher AN content were, irrespective of the type of column used, always eluted later than samples of lower AN content. THF is a solvent for S/AN samples but *n*-hexane is a nonsolvent. The latter statement holds true for *iso*-octane as well.

The chromatograms shown in Fig. 5.8 were obtained through a gradient *i*Oct/THF. The sample (a mixture of five S/AN model copolymers), injection volume, flow rate, and attenuation were identical in each chromatogram, but the packings of the columns were quite different: bare silica, C18 ($d_0 = 6$ nm), C18 ($d_0 = 400$ nm), and μBondagel. The small shift of the time axis in both the latter chromatograms is a consequence of different column geometry which influences the gradient dwell-time in the respective systems. Similar chromatograms (with even more details in the second and the last peak) were obtained on a nitrile bonded-phase column.

The correspondence of the elution patterns suggests separation by solubility effects. This idea is proved by the fact that the NP gradient used (*i*Oct/THF) has the same separating efficiency in combination with RP C18 columns as well as

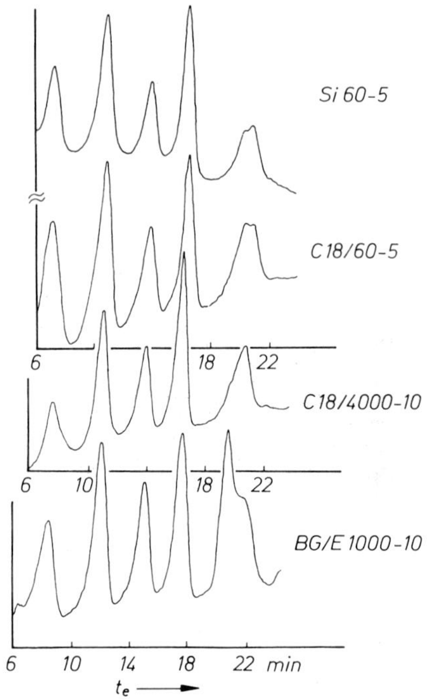

Fig. 5.8. Separation of the mixture of five *stat*-copoly(styrene/acrylonitrile) samples on four different columns through a gradient *iso*-octane/(tetrahydrofuran, with 10% methanol added) [13]. Gradient: 10% **B** at $t = 0$, linearly increased to 50% in 2 min, further by 2.5%/min to 100% **B** at 22 min. Columns "Si60-5": 150 × 4.6 mm, packed with Polygosil 60–5 ($d_0 = 6$ nm, $d_p = 5\,\mu$m); "C18/60–5": 150 × 4.6 mm, packed with Polygosil C18 ($d_0 \leqq 6$ nm, $d_p = 5\,\mu$m); "C18/4000-10": 150 × 4.6 mm, packed with LiChrosorb Si 4000 C18 ($d_0 = 400$ nm, $d_p = 10\,\mu$m); "BG/E 1000–10": 300 × 4.0 mm, packed with μBondagel E1000–10 ($d_0 = 100$ nm, $d_p = 10\,\mu$m). Samples (in the order of elution) A–E (see Table 9.1) with 16, 23, 29, 36, or 43 mass% AN

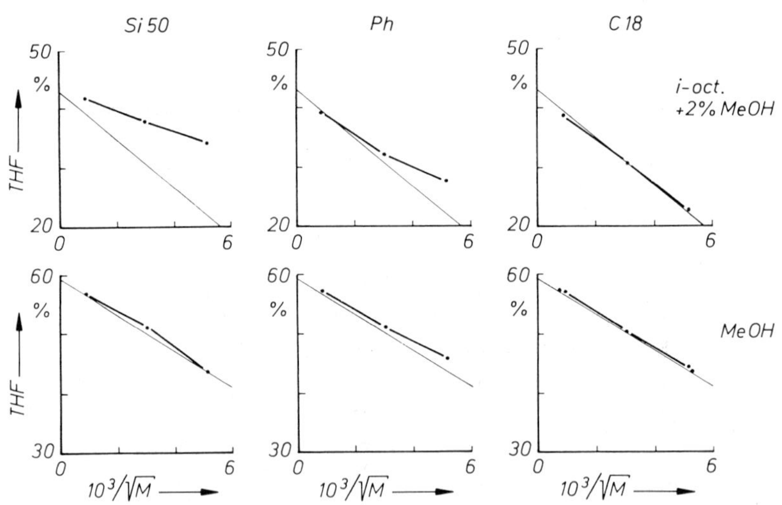

Fig. 5.9. Molecular-weight dependence of solubility and retention of polystyrene standards in two solvent systems [7]. Upper diagrams: *iso*-octane/tetrahydrofuran (with 2% methanol throughout), lower row: methanol/THF. *Light lines* by turbidimetric titration, *dots* and *heavy lines* by gradient elution on three columns, each 60 × 4 mm; ("Si50": Nucleosil Si50, $d_0 = 5$ nm, $d_p = 5\,\mu$m; "Ph": Nucleosil phenyl-bonded-phase, $d_0 \leqq 5$ nm, $d_P = 7\,\mu$m; "C18": nucleosil C18, $d_0 \leqq 5$ nm, $d_p = 5\,\mu$m)

with a silica column, although only the latter combination forms a proper NP system.

Irregular combinations of gradient orientation and column polarity are also effective in the separation of PS samples by MW. Figure 5.9 gives a synoptic presentation of results measured with two gradients on three different columns. The upper row of plots shows the effect of a NP gradient (i.e. increasing in polarity, formed by the addition of THF to *i*Oct, with 2% MeOH throughout). The other gradient was formed by addition of THF to MeOH, i.e. it was a RP gradient decreasing in polarity. The combination of the latter with a C18 column yielded a proper RP system whose characteristics are shown in the lower right-hand diagram in Fig. 5.9. The elution of PS standards in this RP system is indicated by the points and the connecting heavy line. The latter almost follows the light line which indicates the solubility of PS in THF/MeOH mixtures. (This part of Fig. 5.9 is a condensed version of Fig. 5.6.)

A MeOH/THF gradient in combination with a silica Si50 column does not fit into the approved system of chromatographically effective phase combinations, nor does an *i*Oct/THF gradient in combination with a RP C18 column. Nevertheless, these irregular combinations yielded reasonable separation according to MW of PS samples. This must be caused by precipitation/redissolution effects.

A chromatographic efficiency of irregular phase combinations is a rigorous proof of a solubility mechanism. With a mechanism of this kind, the position of the elution characteristics and solubility lines are identical within the limits of experimental precision. In the diagrams of the proper NP system *i*Oct/THF on Si50 and also of the combination of a gradient *i*Oct/THF with a phenyl bonded-phase column, the position of the elution characteristics (above the solubility line) indicates adsorption contributions.

On the other hand, coincidence of elution characteristics and solubility lines suggests (but does not prove) a solubility mechanism. The contribution of adsorption to retention can be so small that it may be not visible in an usual graph but nevertheless essential for efficient separation. This can be deduced from the discussion of stationary phase-effects on separation, see Sect. 6.2.

5.4 Effect of Temperature

Decreasing eluent strength causes transition from SEC to AC behaviour, see Sect. 6.1. In SEC, elution time decreases with increasing MW, in AC it increases. At the critical value of interaction energy, the enthalpy term just compensates for the entropy contribution, and samples of any MW are equally retained.

The balance is limited to a certain temperature. Increasing temperature yields prolonged retention [14], see e.g. line "40" in Fig. 5.10. At a first glance, this seems surprising but it can be understood by taking into account the entropy balance of the whole system: the entropy increase caused by displacing solvent molecules from the surface is greater than the entropy loss due to adsorption of a corresponding number of repeat units which, in polymers, are interlinked and

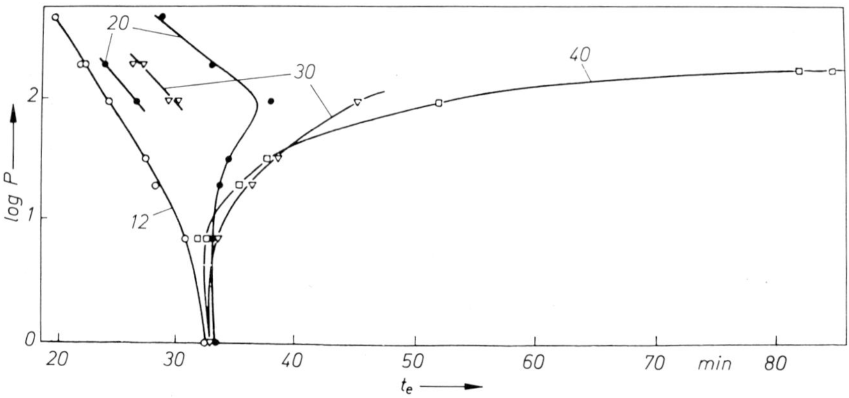

Fig. 5.10. Effect of temperature on the elution of polystyrene standards from a silica column (600×4 mn, $d_0 = 10$ nm). Ordinate: log(degree of polymerization), eluent tetrachloromethane + chloroform (94.5:5.5), isocratic. Column temperature in Centigrade indicated at the curves. From Ref. [14] with permission

thus ordered even before adsorption. Further to this, increasing temperature conformationally favours polymer adsorption.

When retention in non-exclusion HPLC increases with increasing temperature, the underlying mechanism is certainly adsorption. Retention increasing with temperature has been found with S/MMA copolymers (26.6 – 84.8 wt-% MMA) on a silica column by gradient elution using **A** = TCM + EtOH (99:1) and **B** = TCM + EtOH (95.5:4.5) [15]. The flow rate was 0.5 ml/min and the

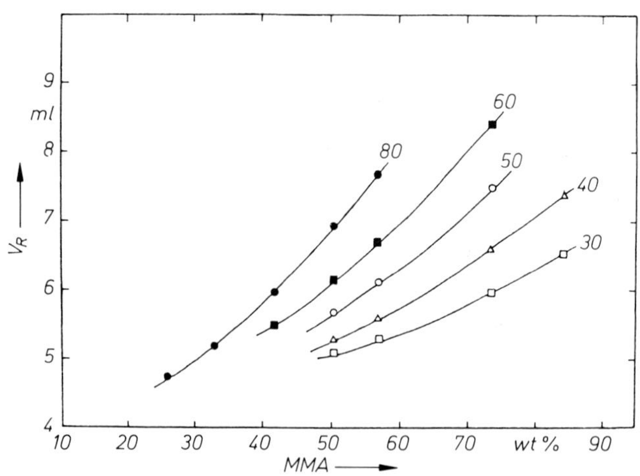

Fig. 5.11. Influence of temperature on the elution characteristics (retention vs sample composition) of *stat*-copoly(styrene/methyl methacrylate) samples. Silica column (50×4.6 mm, $d_0 = 3$ nm, $d_p = 5\,\mu$m), gradient chloroform/ethanol (1–4.5% in 15 min), flow rate 0.5 ml/min, $V_0 = 50\,\mu$l, $m_0 = 25\,\mu$g. Column temperature in Centigrade indicated. From Ref. [15] with permission.

gradient program 0–100% **B** in 15 min. For a copolymer of given composition, retention increased with temperature, see Fig. 5.11.

Transition from zero to infinite retention could also be observed as a result of rising temperature: in isocratic elution, a sample containing 33.7 mol-% MMA left the column in the interstitial volume as a sharp peak of constant height at temperatures below 30 °C but remained on the column at 60 °C. In the transition region, the peak height decreased from 100% (25 °C), to 85% (30 °C), 48% (40 °C), and 6% (50 °C). The retention volume remained unchanged [16]. Similar behaviour was observed with S/MA, S/EA, S/BA, S/EMA, and S/BMA copolymers [17].

On the other hand, decreasing retention with increasing temperature points towards a solubility mechanism because polymer solubility increases with temperature in almost any case.

At any rate, whatever the retention mechanism might be, polymer HPLC is strongly influenced by temperature. Thus, control of column temperature is indispensable. Temperatures as high as the eluent system permits should be chosen in order to facilitate diffusion and retention kinetics.

5.5 High-Performance Precipitation Liquid Chromatography

The discussions in the antecedent sections are based upon the perception that usually "solubility and adsorption effects are operative in polymer HPLC. Sometimes the former prevail, sometimes the latter are indispensable for good separations" [18]. Best chances for successful gradient chromatography can be expected by synergic combinations of adsorption and precipitation effects.

With stationary phases of high activity, proper adsorption chromatography is possible with polymer samples whose structure allows complete elution in the course of a gradient, i.e. elution without any irreversibly retained portions of an injected sample. The term proper adsorption chromatography refers to regular combinations of stationary and mobile phases (see Sect. 6.2) and samples dissolved in the starting eluent.

If the elution strength of available solvents does not suffice for complete elution a less active column must be chosen. With moderate interaction forces, elution will certainly be no longer the problem but retention may become difficult (see Sect. 8.3). In this case, the starting eluent must be rather poor in order to facilitate retention. Since the solubility window of a polymer is relatively narrow, a poor eluent is often a nonsolvent at the same time. Then it is impossible to prepare sample solutions in the starting eluent whose quality is determined by the condition of retaining the polymer in a column. Thus, synthetic polymers are often chromatographed as solutions in solvents which are thermodynamically better than the starting eluent. Consequently, the polymer will be precipitated on injection.

The term high-performance precipitation liquid chromatography (HPPLC) has been suggested for gradient elution with sample injection into a starting eluent which is a nonsolvent (precipitant) for the sample under investigation

[11, 13]. This definition does not at all deny the contribution of adsorption to retention and separation. In the contrary, just the hypothesis of synergic action of precipitation and adsorption proved a guiding principle in the search for new separation methods [19–21]. It showed itself reliable even for developing normal- and reversed-phase HPLC of S/EMA mixtures with reversal of elution order [22].

It should be noted here that protein separations are usually performed in a different way, i.e. with water as sample solvent and starting eluent. However, samples of poor solubility, e.g. membrane proteins or collagen, are injected as solutions in liquids thermodynamically better than the starting eluent. In separations of this kind, intermediate precipitation is also observed.

Method evaluation in HPPLC is most efficiently performed with stable solutions of standard samples in a pure solvent, which should be a component of the gradient mixture. Retention via precipitation requires exchange of the good solvent in the volume injected against a precipitating mixture. In order to complete this exchange as soon as possible and to retain the sample already at the top of the column, the starting eluent must be rather poor. As a consequence of this, the elution of a retained sample may occur only after a corresponding increase in eluting strength, i.e. with a considerable delay.

A gap between retention and elution conditions can be understood also for an adsorption mechanism on base of the model of multiple attachment (see Sect. 4.3). Adsorption from the mobile phase requires k'_u to be large enough for immobilizing a solute via a single monomeric unit even against the mechanical counteractions of the streaming liquid. As soon as a flexible polymer is immobilized, it will exhibit more and more interactions with the stationary phase until it is adsorbed via extended trains of monomeric units. In this state of adsorption, the *capacity factor* reaches, according to Eq. (4.14), a high value k'_{total} based on the k'_u value necessary for catching the polymer solute. Elution requires a considerable reduction of k'_u.

The transition in eluent composition from the conditions needed for retention to the composition suited for elution can be accelerated by a high initial gradient rate but the elution of the sample should be carried out at a moderate gradient rate. Thus, multilinear gradients are often advantageous in HPPLC.

Different gradient rates have influence on elution time, of course, but the eluent composition at peak elution is not altered. For calculation of the latter, precise knowledge of the gradient *dwell time* is indispensable. Experimental evidence for equivalence of different gradients in the elution of S/MMA copolymers through Hex/TCM gradients has been reported [23].

5.6 Normal-Phase Gradient Elution with Separate Adjustment of Polarity and Solvent Strength

Under conventional working conditions, normal phase HPLC of polymers is carried out by adding a polar solvent to a nonpolar liquid, which is often a non-solvent for the polymer samples to be investigated. In this section, report is given

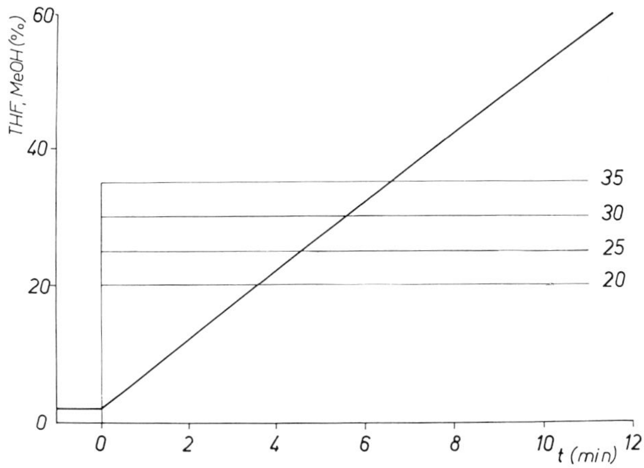

Fig. 5.12. Sudden-transition gradient schemes: methanol concentration rising by 5%/min (heavy line) after a sudden increase of tetrahydrofuran content from zero to the indicated level (20, 25, 30, or 35 vol%, light lines); the third component of the eluent is *iso*-octane)

on attempts to eliminate the solubility influence and elute the sample components mainly by increasing the polarity of the eluent. For this purpose, eluent mixtures of *i*Oct, THF, and MeOH were applied on CN bonded-phase columns. In order to ensure proper retention, the sample solutions in THF were injected into *i*Oct + 2% MeOH as a starting eluent. (As already mentioned in Sect. 5.2.1 the MeOH addition was used for supressing ghost peaks.)

In polarity gradients after a sudden increase in solvent concentration ("sudden-transition gradients"), the concentration of THF was, simultaneously to sample injection, raised from zero to 20% or more and held constant at that level. The elution itself was carried out by the addition of 5% MeOH per min. The scheme of gradients of that kind are presented in Fig. 5.12. Of course, the sudden increase in THF content caused a disturbance of the elution pattern, but one minute after the transition the system behaved steadily again. With copolymers of S and methacrylate esters (MMA, EMA, MEMA) and a suitable level of THF concentration, elution could be achieved solely by increasing the MeOH content of the eluent.

Figure 5.13 shows the elution patterns of five S/EMA copolymers through *i*Oct/MeOH gradients at constant THF content of 35, 30, 25, or 20% (from top to bottom). The elution time increased with EMA content of the samples (A < C < E < G < I, see Table 9.5). At 20% THF concentration, elution from the CN column solely by increasing the MeOH concentration was difficult. This can be seen by the rather long chromatogram and the (comparatively) bad peak shape. Best results were obtained with 25 or 30% THF content. For S/EMA copolymers, 35% THF was too high an addition, since specimen A (4.7% EMA) was not properly retained. Presumably, it was washed out during the sudden increase in THF concentration. The poor peaks in the lowermost recordings with 20% THF

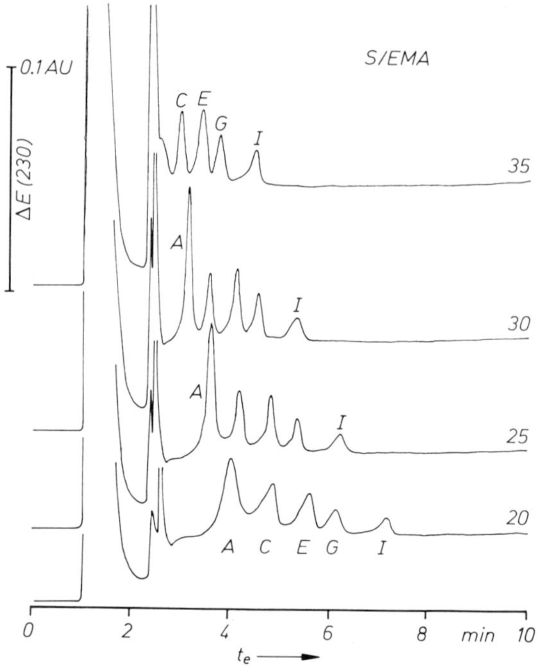

Fig. 5.13. Separation of the mixture of five *stat*-copoly(styrene/ethyl methacrylate) specimens through sudden-transition gradients *iso*-octane/methanol at the THF concentrations indicated at the curves. Column 60 × 4 mm, packed with CN bonded-phase, $d_0 \leqq 5$ nm; $d_p = 5\,\mu$m. Flow rate 0.5 ml/min, 50 °C. Sample: 1.8 μg copolymer A (4.7 mass% EMA, $10^{-3} M_w = 51.6$) + 1.2 μg C (32.2%, 63.1) + 2.0 μg E (54.6%, 65.2) + 1.2 μg G (68.0%, 83.6) + 2.0 μg I (92.5%, 61.6) in 20 μl THF. UV signal at 230 nm

indicate that MeOH is not in any respect an adequate substitute for THF. Both liquids are only partially equivalent.

In efforts to adjust polarity and solvent strength separately one must be aware of the co-solvency effect of MeOH, see Sect. 5.2.1. The badly shaped elution patterns at 20% THF concentration in Fig. 5.13 indicate that, at higher values of MeOH concentration, the co-solvency effect is rather small. Thus, the main result of increasing MeOH concentration is indeed the gain in polarity.

Figures 5.14 and 5.15 present analogous results for S/MMA and S/MEMA copolymers. They correspond to the results shown in Fig. 5.13 but the THF level for optimum separation was 35% with S/MMA and 30 or 35% with S/MEMA samples. These copolymers are more polar than corresponding S/EMA samples and, therefore, more strongly retained on a CN bonded-phase column. It is surprising that the THF content must be raised in order to overcome the stronger retention, i.e. even in the case of strong adsorption interactions a rather high level of solvent strength is required for a neat separation; the higher adsorption contribution cannot sufficiently be compensated by MeOH alone.

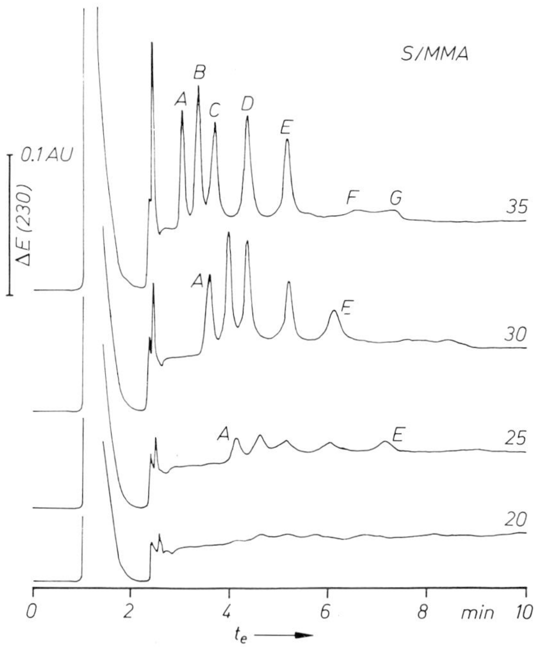

Fig. 5.14. Separation of the mixture of seven *stat*-copoly(styrene/methyl methacrylate) specimens through sudden-transition gradients, conditions see Fig. 5.13. Sample: 1.6 μg A (16.6 mass% MMA, $10^{-3}M_{OSM} = 120$) + 1.8μg B (21.4%, 133) + 1.6 μg C (30.1%, 133) + 1.6 μg D (52.7%, 160) + 1.8 μg E (58.6%, 160) + 1.6 μg F (71.4%, 55.5) + 1.8 μg G (84.8%, 86) in 20 μl THF

A triangle diagram of the three-component eluent *i*Oct/THF/MeOH enables graphical representation of the elution characteristics. The hatched area in Fig. 5.16 marks the range where elution was observed. Elution characteristics running parallel to the hypotenuse (THF–*i*Oct) of the triangle would indicate elution by a constant MeOH concentration.

Figure 5.17 repeats the hatched section of Fig. 5.16 together with the elution characteristics of the copolymers investigated. The line "MeOH=0" is the hypotenuse. None of the measured characteristics is parallel to this line, i.e. elution at a constant MeOH concentration was not observed. This can be understood by the fact that addition of THF facilitates solubility as well as increases the polarity of the eluent. Thus, at a higher level of THF concentration, less MeOH is required for elution.

The more the directions of the elution characteristics approaches a vertical orientation the better is MeOH suited a substitute for THF. With a vertical position of the elution characteristics (see, e.g. the line for S/MMA sample "E" or, approximately, S/MEMA sample "F"), elution takes place at constant *i*Oct concentration of the eluent. Here, a given volume of MeOH has the same elution power as a corresponding volume of THF. The approach towards vertical

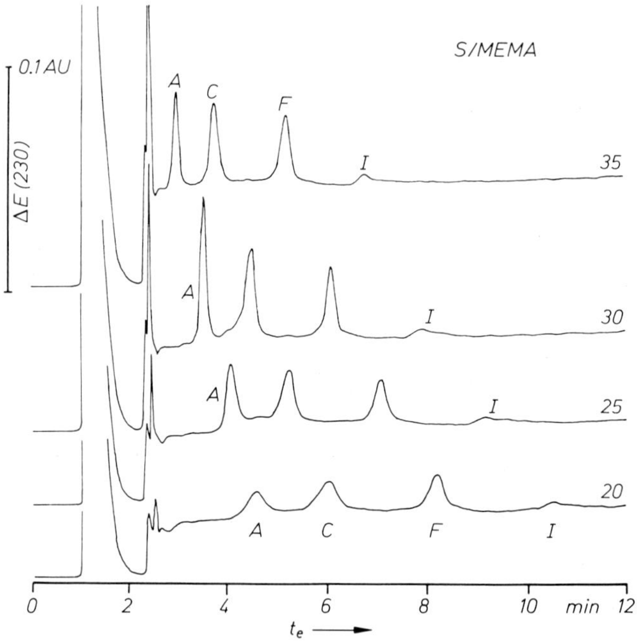

Fig. 5.15. Separation of the mixture of four *stat*-copoly(styrene/2-methoxyethyl methacrylate) specimens through sudden-transition gradients, conditions see Fig. 5.13. Sample: 2.0 μg A (13.4 mass% MEMA, $10^{-3}M_w = 88$) + 2.2μg C 38.0%, 180) + 2.2 μg F (62.4%, 173) + 1.8 μg I (87.4%, 306) in 20μl THF

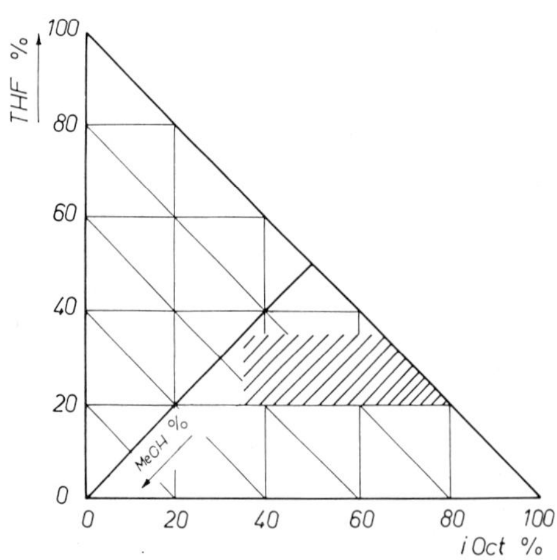

Fig. 5.16. Composition diagram of three-component system *iso*-octane/tetrahydrofuran/methanol. The *shaded area* (20–35% THF, 2–45% MeOH) has been used in the separations shown in Figs. 5.13–15 and is redrawn in Fig 5.17.

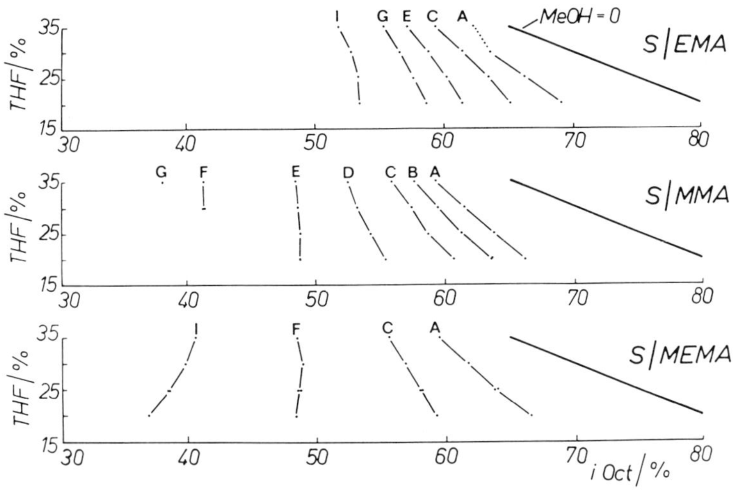

Fig. 5.17. Elution characteristics of the copolymers investigated in Fig. 5.13–15 by sudden-transition gradients; tetrahydrofuran concentration 20, 25, 30, or 35 vol%; methanol concentration in vol% at each point is (100–*i*Oct%–THF%)

orientation becomes closer with increasing content in methacrylate units. Since MMA or MEMA groups increase the polarity of the copolymers, the tendency expressed in Fig. 5.17 can be understood as follows: the higher the polarity of a sample, the larger is the contribution of adsorption to retention and the higher is the eluting power of MeOH relative to that of THF.

The application of gradients with a sudden transition to a level of THF concentration which, by itself, does not suffice for elution, in combination with the subsequent continuous addition of a highly polar displacer with low dissolution power has several advantages.

1. With MeOH as a displacer, the elution can be monitored at comparatively short wavelengths without baseline slope. A constant THF concentration yields a constant shift of the baseline without disturbing the shape of peaks.
2. The almost independent selection of solubility and polarity enables fine-tuning of retention and elution.
3. The divergence of elution characteristics as displayed in Fig. 5.17 provides the chance of increasing the selectivity of HPLC separation. A comparison of Fig. 5.13 with, e.g. Fig. 6.6 illustrates this effect. In both cases, mixtures of unfractionate S/EMA copolymers were investigated by a gradient increasing in a polar component by 5%/min. (The use of a Si50 column in Fig. 6.6 but a CN bonded-phase in 5.13 does not disturb a comparison. This can be concluded from Figs. 6.7 and 6.8 which reveal similarity of Si50 and CN bonded-phase columns in the separation of S/EMA copolymers.)
4. The possibility of choosing shorter wavelengths provides the chance of monitoring sample components which have only a small signal at, say,

259 nm. This is, e.g. important for S/MEMA copolymers. Figure 9.21 shows NP elution by a common *i*Oct/THF gradient and detection at 259 nm. The lowermost trace is from a sample containing 79.7 mass% MEMA. The recording is badly shaped and almost too small for evaluation. Figure 9.22 shows the chromatogram of the mixture of four S/MEMA samples, which was obtained in a sudden-transition gradient and monitored at 230 nm. The sample mixture contained 26.8% of a copolymer with 62.4% MEMA (third peak) and 22.0% of a copolymer with 87.4% MEMA (last peak). The improvement is obvious. (Under the condition used in Fig. 9.21, the last sample could not be detected at all.)

5.7 References

1. Quarry MA, Stadalius MA, Mourey TH, Snyder LR (1986) J Chromatogr 358: 1
2. Glöckner G (1964) Inaugural dissertation, Dresden University of Technology
3. Glöckner G (1983) Pure Appl Chem 55: 1553
4. Engelhardt H, Czok M, Schultz R, Schweinheim E (1988) J Chromatogr 458: 79
5. Schultz R (1989) Thesis, University of Saarbrücken, p. 75
6. Glöckner G (1965) Z Phys Chem (Leipzig) 229: 98
7. Glöckner G (1988) Chromatographia 25: 854
8. Glöckner (1987) Chromatographia 23: 517
9. Glöckner G, Francuskiewicz F, Müller K-D (1971) Plaste u Kautschuk 18: 654
10. Glöckner G, Francuskiewicz F, Müller S (1975) Faserforschg Textiltechnik 26: 287
11. Glöckner G, Kroschwitz H, Meißner Ch (1982) Acta Polymerica 33: 614
12. Glöckner G, van den Berg JHM (1984) Chromatographia 19: 55
13. Glöckner G, van den Berg JHM (1987) J Chromatogr 384: 135
14. Tennikov MB, Nefedov PP, Lagareva MA, Frenkel SJa (1977) Vysokomol Soed Ser A 19: 657
15. Mori S, Uno Y (1987) J Appl Polym Sci 34: 2689
16. Mori S, Uno Y (1987) Anal Chem 59: 90
17. Mori S, Mouri M (1989) Anal Chem 61: 2171
18. Glöckner G (1989) In: Barth HG (ed) Proc First International Symp Polym Analysis and Characterization, Toronto, 1988, J Appl Polym Sci, Appl Polym Symposia 43: 39
19. Glöckner G, Stickler M Wunderlich W (1987) Fresenius Z Anal Chem 328: 76
20. Glöckner G, Stickler M, Wunderlich W (1989) J Appl Polym Sci 37: 3147
21. Glöckner G, Müller AHE (1989) J Appl Polym Sci 38: 1761
22. Glöckner G (1987) J Chromatogr 403: 280
23. Sato H, Takeuchi H, Tanaka Y (1986) Macromolecules 19: 2613

6 Effect of Mobile and Stationary Phase in Polymer HPLC

6.1 Elution in the Critical Region

The multi-site attachment of polymers and the high values of *solute acceleration factor S* (see Eq. (4.21) have the consequence that small differences in eluent composition can cause dramatic changes in elution behaviour, see Sect. 4.3. For example, a polystyrene standard ($M = 50{,}000$) was eluted without retention from a 30 nm RP C18 column by water + THF (13:87 v/v) whereas, with a 17:83 v/v mixture, no elution was observed within reasonable time [1], see Fig. 6.1.

The sudden change in elution behaviour by small variations of solvent strength can occur as a side effect in SEC with mixed eluents. Fig. 6.2a and b show observations of that kind [2, 3]. In common SEC, elution time increases with decreasing MW of the sample, see line 100 in Fig. 6.2a, which shows the behaviour of PS in pure chloroform (TCM). Small changes apart, the S-shaped characteristic is kept on addition of carbontetrachloride, provided the TCM content in the mixture is not less than 5.9%. At 30 °C, the critical concentration is 5.5% TCM and at 27 °C it is 5.35%. At lower TCM content, the macromolecules are retained on the surface of the silica. Since the gain in enthalpy increases with the number of repeat units involved, retention now increases with MW, see lines "5" or "0" in 6.2a and "5.2", "5.1", and "5.0" in 6.2b.

Similar results have been found in related work with PS standards on silica in Hex/DCM mixtures ($c_{crit} = 59\%$ DCM) [4] or decalin/THF ($c_{crit} = 23\%$ THF) [5], with PS standards on Partisil "PAC" in *i*Oct/DCM ($c_{crit} = 45\%$ DCM) [6], with PS standards on μBondagel in Hp/THF ($c_{crit} = 40\%$ THF) [7], with poly(butylene terephthalate) oligomers on silica in Hp/THF ($c_{crit} = 65\%$ THF) [3], or with butadiene oligomers on silica in Hp/MEK ($c_{crit} = 0.5\%$ MEK) [3].

At critical eluent composition, the retention is (in a certain range) almost independent of MW. Isocratic elution in the critical region is capable of separating by functionality without superimposition of MW effects [3, 8, 9], see Fig. 6.3. The method has been used, e.g. for the separation of telechelic polymers [8]. A survey on critical elution and its applications is given in Ref. [3].

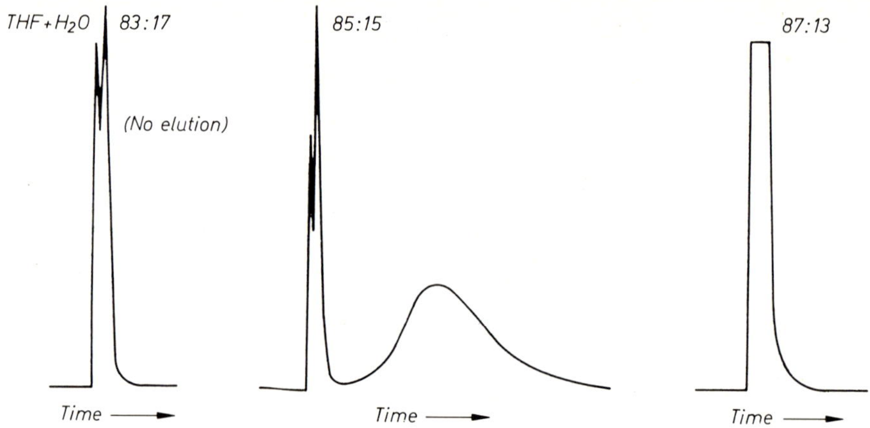

Fig. 6.1. Isocratic elution of a polystyrene sample in THF + water mixtures. Silica column, $d_0 = 30$ nm. Sample dissolved in THF, eluent composition indicated. At 17% W in the eluent, only the sample solvent passes the UV detector, the polymer remains on the column. From Ref. [1] with permission

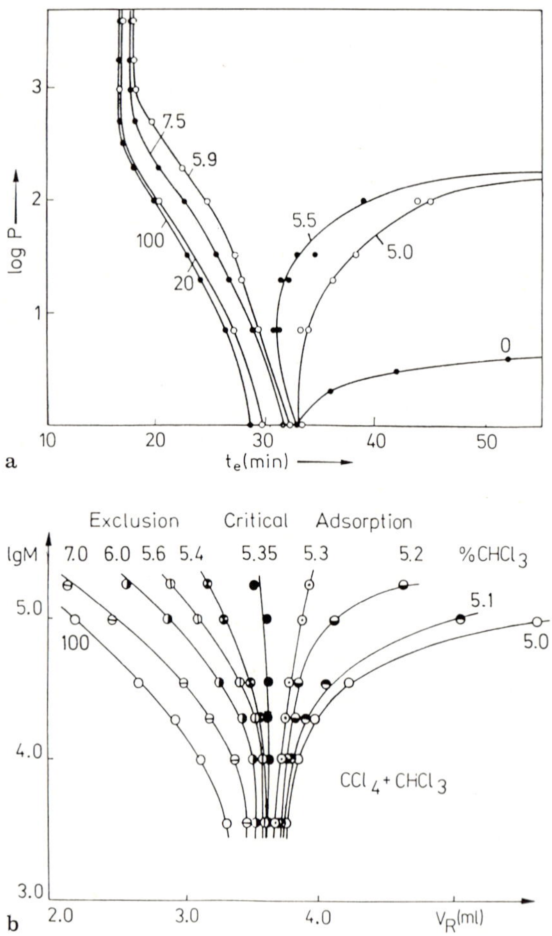

Fig. 6.2 a, b. Solvent-strength effect on the elution of polystyrene standards from silica columns. Eluent: tetrachloromethane + chloroform mixtures, vol% chloroform indicated. (**a**) 30 °C, $d_0 = 10$ nm; from Ref. [2] with permission; (**b**) 27 °C, $d_0 = 30$ nm; from Ref. [3] with permission

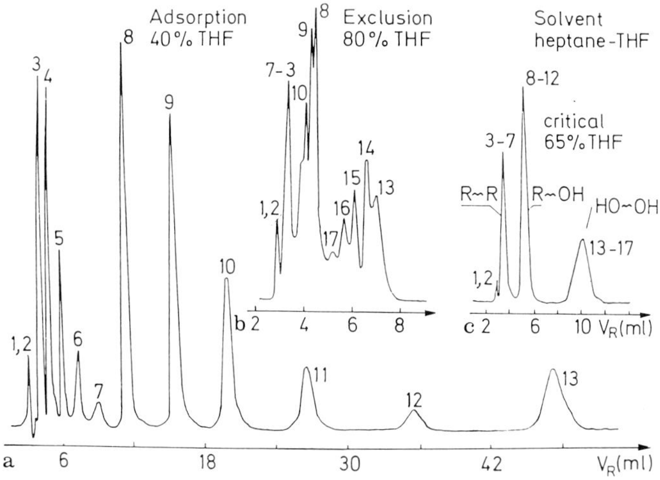

Fig. 6.3. Isocratic separation of oligomers through heptane + tetrahydrofuran mixtures with either 80%, 65%, or 40%, THF content; (**a**) adsorption HPLC, (**b**) SEC separation, (**c**) elution in the critical region: group separation. Sample: poly(butylene terephthalate) oligomer mixture with $P = 0$–4 repeat units ($P = 0$: 1,4-butane diol) and f = 0, 1, or 2 terminating OH groups. Peaks 3–7: f = 0, $P = 0$–4; peaks 8–12: f = 1, $P = 0$–4; peaks 13–17: f = 2, $P = 0$–4; (1 and 2: solvent peaks). Silica column (d_0 = 6 nm). 24 °C, flow rate 1 ml/min, $V_0 = 10\,\mu$l. UV detection at 254 nm. From Ref. [3] with permission

6.2 The Role of the Stationary Phase

In HPLC, column packings must stand high pressure and, in gradient HPLC, the grain size must not change when the eluent composition varies. Therefore, packings of modified (or bare) silica are usually preferred but Sato et al. [10] performed gradient HPLC also on crosslinked poly(acrylonitrile) or polystyrene and found these packings suited for gradient HPLC, provided that the materials had been polymerized from starting mixtures containing 33–36 vol% of a crosslinking monomer.

The diagrams 1 and 2 in the upper row of Fig. 5.9 indicate that, in a gradient *i*Oct/THF, PS samples are eluted in the field of stable solutions if silica or phenyl bonded-phase columns are used. Here, the retention is influenced by adsorption effects.

In general, the approved rules of phase combination in HPLC are valid also in polymer HPLC. Exceptions are the elution of S/AN specimens by gradients *i*Oct/THF, which indicated almost identical behaviour on different columns, see Fig. 5.8, or of PS in certain phase combinations, see Fig. 5.9, diagrams 3–6. In these cases, the type of column is almost without effect on the elution sequence of samples and the separation of mixtures.

Adherence to the rules of phase combination can be observed in the elution of S/MMA or S/EMA copolymers in NP gradients *i*Oct/THF. Figure 6.4 shows the elution patterns of S/EMA copolymers by a gradient *i*Oct/THF on a C18 column. The specimens are eluted in a period ranging from approximately 5 to

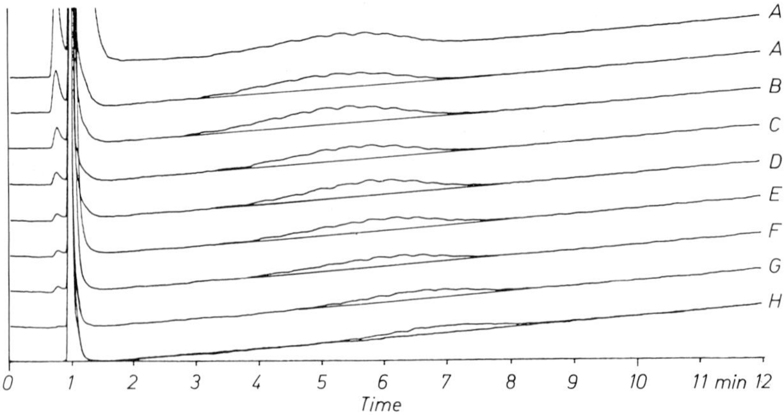

Fig. 6.4. Gradient elution of *stat*-copoly(styrene/ethyl methacrylate) samples on a C18 column through *iso*-octane/tetrahydrofuran (20–80% in 12 min) [13]. Column 60 × 4 mm; $d_0 \leqq 5$ nm, $d_p = 5$ μm; 50 °C, flow rate 0.5 ml/min; sample data: see Table 9.5. Injection: 5 μl each, containing 5 μg of the respective copolymer in THF (the uppermost curve is from 10 μl). UV signal at 259 nm, attenuation 50 mAU, experimental baseline plotted together with the elution curve of sample "H" and redrawn with the others

7 min. Although all individual patterns are rather broad, they are related to sample composition (see Table 9.5) in a logical order: the higher the EMA content the later is a sample eluted, i.e. it is delivered in an eluent richer in THF. The same sequence was observed in TT of S/EMA copolymers in THF with *n*-hexane (Hex), i.e. the higher the EMA content of a sample the less Hex nonsolvent is required for reaching the cloud point.

The phase combination used for the experiments in Fig. 6.4 was an irregular one. A proper system can be formed by combining the NP gradient *i*Oct/THF

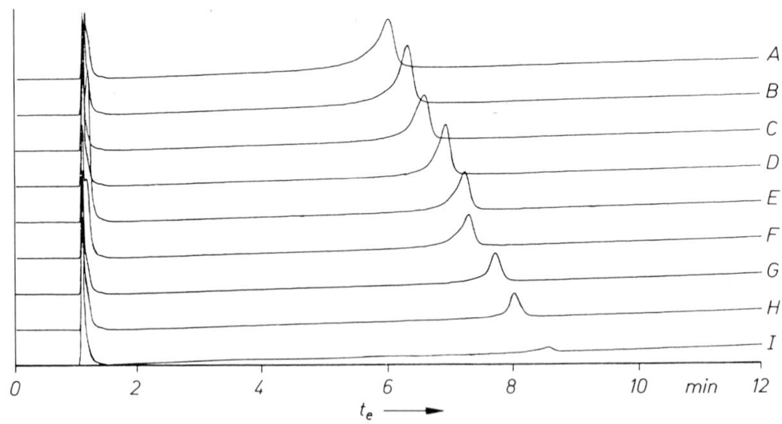

Fig. 6.5. Repetition of the experiments shown in Fig. 6.4 on a CN bonded-phase column (60 × 4 mm; $d_o \leqq 5$ nm, $d_P = 5$ μm), all other experimental condition identical. Attenuation 150 mAU

with a polar column. The result of this alteration is shown in Fig. 6.5. Samples and gradient were the same as in Fig. 6.4 but a CN bonded phase column was used instead of a C18 column. The improvement in peak shape is obvious. It must be emphasized that column dimensions, column temperature, flow rate, and the grain size of the packings were the same in both sets of experiments. Thus, the badly shaped elution patterns in Fig. 6.4 were not caused by ordinary peak broadening.

Figure 6.6 shows the NP separation of the mixture of four S/EMA specimens on a silica column through the same gradient *i*Oct/THF as used in Figs. 6.4 and 6.5. The copolymers were rather low in MW. Thus, the separation by composition was superimposed by a MW effect which, according to Eq. (5.3), is the more pronounced the smaller the MW of a polymer is. Figure 6.7 shows the elution of S/EMA copolymers through a similar gradient but with prefractionation by SEC. Now, the components of the mixture are almost baseline separated.

The same mixture of five S/EMA copolymers was also investigated on a CN bonded phase column of identical geometry. The gradient was the same as well. The result shown in Fig. 6.8 fully corresponds to that obtained at a Si50 column.

Copolymers of styrene and methyl methacrylate are likewise soluble in THF and insoluble in *i*Oct. The content in MMA units determines the THF concentration needed for a stable solution. The solubility borderline can be estimated by TT of solutions in THF through controlled addition of a precipitant. The less MMA a sample contains the more precipitant is needed for segregation of gel droplets. In Fig. 6.9, line "b" represents the turbidimetric

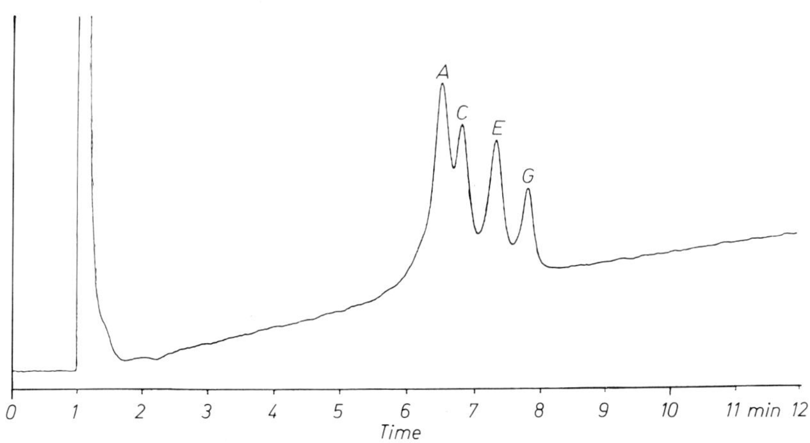

Fig. 6.6. Normal-phase separation of the mixture of four *stat*-copoly(styrene/ethyl methacrylate) specimens. [13]. Silica column (60 × 4mm; $d_O = 5$nm, $d_P = 5\mu$m), gradient and chromatographic conditions as in Figs. 6.4 and 6.5. Sample: 2.5 μg each of S/EMA copolymers A, C, E, and G (copolymer data see Table 9.5) in 10 μl THF, 25 mAU full scale

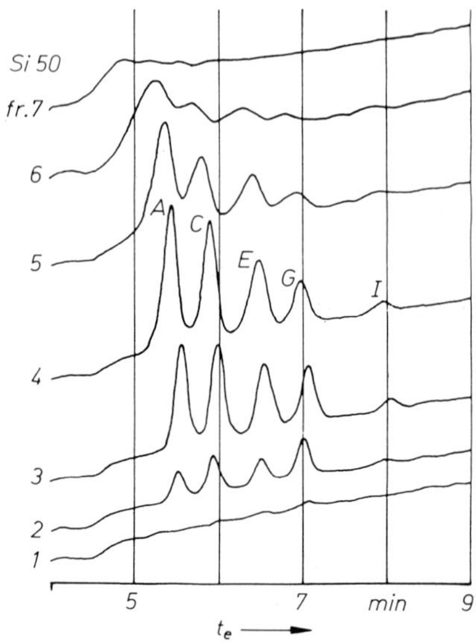

Fig. 6.7. Normal-phase gradient separation of the mixture of five *stat*-copoly(styrene/ethyl methacrylate) specimens after prefractionation by SEC [18]. Injection volume in gradient HPLC 100 μl SEC eluate. Silica column as in Fig. 6.6, gradient *iso*-octane/tetrahydrofuran, multilinear (0% ($t = 0$), 30% (1 min), 70% (9 min), with 1% methanol throughout). Flow rate 0.5 ml/min, 50 °C, UV signal at 259 nm, chromatograms redrawn from plots of 25 mAU full scale. Starting sample for SEC fractionation contained 19.6% A, 19.8% C, 20.5% E, 19.6% G, and 20.5% I (copolymer data see Table 9.5)

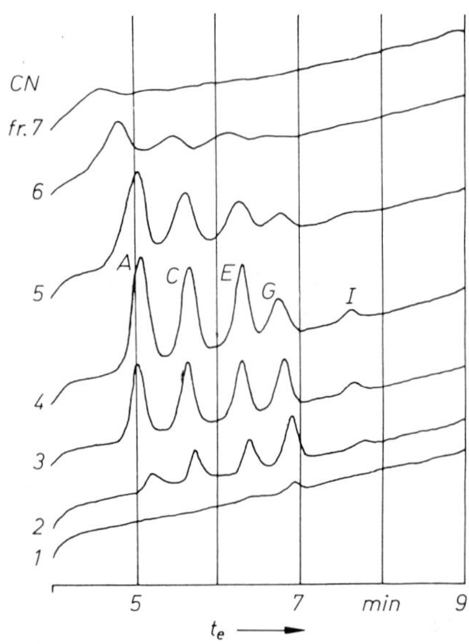

Fig. 6.8. Repetition of the separation shown in Fig. 6.7 on a CN bonded-phase column (60 × 4 nm; $d_o \leq 5$ nm, $d_P = 5$ μm) under otherwise identical conditions [18].

results of S/MMA copolymers in a THF/Hex system (which is almost equivalent to THF/*i*Oct).

Figure 6.10 shows that the retention of S/MMA samples on a RP column in a NP gradient increases with MMA content but the elution curves are rather broad. Curve "d" is the recording obtained on injection of the mixture of the three samples used separately in experiments "a", "b", and "c". The curve of the mixture

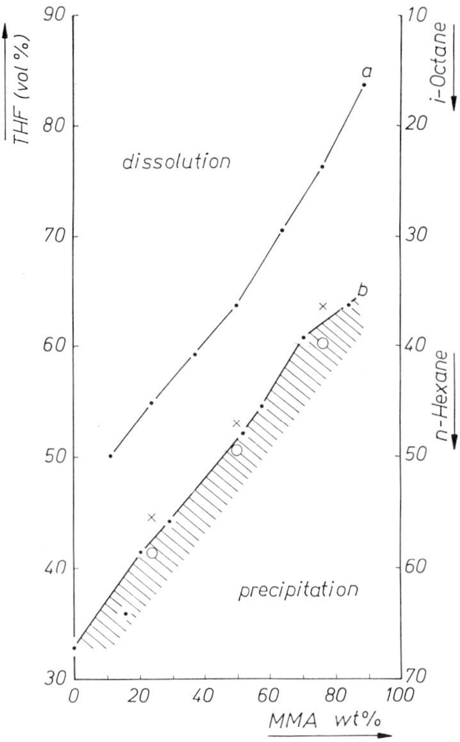

Fig. 6.9. Elution characteristics in *iso*-octane/tetrahydrofuran and solubility of *stat*-copoly(styrene/methyl methacrylate) samples. o: gradient elution on RP C18 column, see Fig. 6.10 x: the same, but on silica column, Fig. 6.11 *a*: elution characteristics of the separation shown in Fig. 9.3 (seven S/MMA standards on silica column, 150 × 4.6 mm, $d_o = 6$ nm, $d_P = 5\ \mu$m) *b*: solubility borderline, estimated by turdimetric titration of S/MMA solutions in THF with *n*-hexane as a precipitant, 20 °C

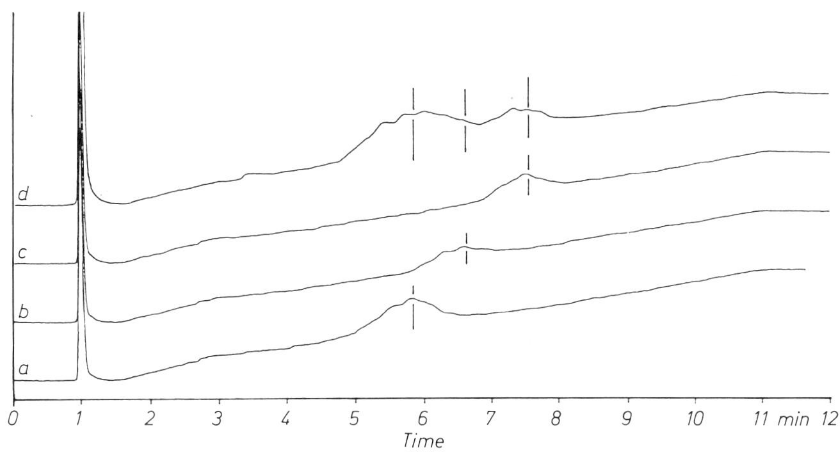

Fig. 6.10. Gradient elution of *stat*-copoly (styrene/methyl methacrylate) samples on a C18 column through a gradient *iso*-octane/tetrahydrofuran (5–100% in 9.5 min). Column 60 × 4 mm; $d_O \leqq 5$ nm, $d_P = 5\ \mu$m; 50 °C, flow rate 0.5 ml/min. Samples: "*a*": 2 μg S/MMA copolymer B, "*b*": 3.3 μg D, "*c*" 5.8 μg F (sample data see Table 9.3); $V_O = 3\ \mu$l each, solvent THF; "*d*": 9μl-injection of a mixture of equal portions of the solutions used in the injections *a–c*. UV signal 259 nm, 50 mAU full scale, first moment of the peaks visible in tracings *a–c* indicated

can scarcely be called a chromatogram. Thus, in contrast to S/AN copolymers, S/MMA samples cannot be separated according to composition by using the combination of a RP C18 column and a NP gradient *i*Oct/THF.

Figure 6.11 repeats the elution curve "d" of the mixture from Fig. 6.10 with the indication of the first moments of the three constituents. Above this curve is the chromatogram given which was obtained with the same sample in the same gradient, but on a silica column [11]. All other chromatographic parameters were identical, including the brands of liquids, the supplier of the columns, and the silica base material for both packings. The separation on the silica column was performed just one hour later than the corresponding injection on the C18 column. A closer look at Fig. 6.11 shows that the three copolymers are eluted from a silica column at a somewhat higher THF concentration than from a C18 column.

The THF concentration at the first moments of the broad patterns in Fig. 6.10 is represented in Fig. 6.9 by circles, whereas the position of the neat peaks from Fig. 6.11 is shown by crosses. A stronger retention on a silica column can bee seen which indicates a higher activity of the latter as compared with a C18 RP column. (Note that the crosses are only slightly above the solubility borderline although adsorption is crucial for separation in this case.)

Curve "a" in Fig. 6.9 indicates the elution characteristics obtained in the separation of the mixture of seven S/MMA copolymers on another silica column, see Fig. 9.3 [12]. All elution characteristics of S/MMA separations on silica are in the range of homogeneous solutions. The distance to the solubility borderline indicates the contribution of adsorption to retention.

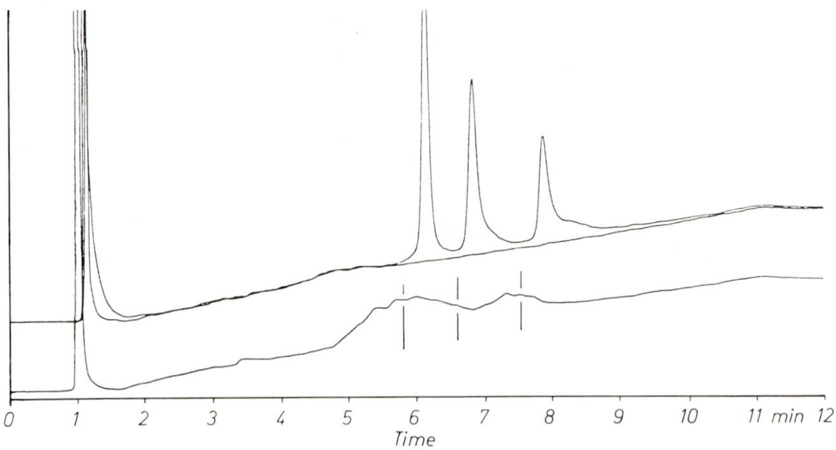

Fig. 6.11. Repetition of injection "*d*" on a silica column (60 × 4 mm; $d_O = 5$ nm; $d_P = 5$ μm) under otherwise identical conditions. (The lower tracing is a copy of the elution curve "*d*" from Fig. 6.10.) [11]

6.3 Normal and Reversed Phase HPLC of Polymers

Gradient elution in NP mode requires the employment of a polar column together with a gradient whose polarity increases in the course of the run. In a proper NP system, retention increases with sample polarity. The elution sequences of S/MMA on silica are in agreement with the general rule in NP chromatography: a higher MMA content increases the polarity of a copolymer. Thus, specimens of higher MMA content are eluted later than others.

Gradient elution in RP mode requires the use of a nonpolar column together with a gradient whose polarity decreases in the course of the run. In a proper RP system, retention decreases with sample polarity. Copolymers of high content in polar units should elute before others. For example, with S/EMA copolymers, the sample with highest EMA content should elute first in RP mode but last in NP mode.

The inversion of elution order is among the most important features of NP vs RP separations. It was a challenging question in the comparison of low MW and polymer HPLC, whether or not the elution order could be inverted also in polymer HPLC. With S/EMA copolymers it was demonstrated that mixed samples can be eluted in sequences increasing as well as decreasing in EMA content [13].

Figures 6.6 and 6.7 have already shown the NP separation of mixtures of S/EMA specimens on a silica column. In Fig. 6.12, the solubility line of S/EMA copolymers in THF is plotted as percentage THF vs w_{EMA}. It was estimated by TT with *n*-hexane. The elution characteristics in the NP system Si50 column plus *i*Oct/THF gradient have been measured with fraction No. 4 of Fig. 6.7 and

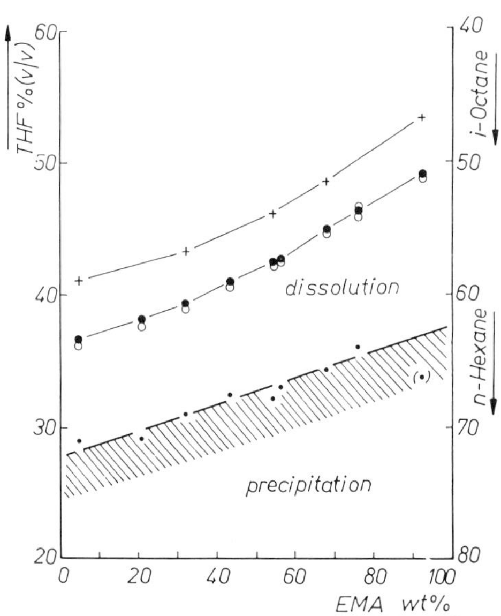

Fig. 6.12. Elution characteristics in *iso*-octane/tetrahydrofuran gradients and solubility of *stat*-copoly (styrene/ethyl methacrylate) samples. +: Gradient elution on silica column, data from Fig. 6.7, fraction No. 4, o, •: elution characteristic on two CN bonded-phase columns of same geometry (60 × 4 mm; $d_O \leqq 5$ nm: $d_P = 5\ \mu m$). Solubility borderline measured by turbidimetric titration (*dots*) in THF with *n*-hexane at 20 °C.

are indicated by upright crosses and the respective connecting line. Figure 6.12 displays also the elution characteristics of S/EMA copolymers in an *i*Oct/THF gradient on a nitrile bonded-phase column (see Fig. 6.5), indicated by full and open circles for two sets of experiments. Both characteristics lie in the range of homogeneous solutions. Obviously, retention is stronger on silica than on nitrile bonded-phase, which can be understood as a consequence of the higher polarity of silica.

Figure 6.13 demonstrates that the separation of S/EMA copolymers can be also achieved in a RP mode: the mixture of five S/EMA samples is separated by composition on a C18 column through a gradient MeOH/THF. The elution order is inverted. As discussed with Fig. 6.6, the separation is superimposed by a MW effect which can be eliminated to a good deal by SEC prefractionation, see Fig. 6.14. (This observation and also the results presented in Fig. 5.9 cannot support the statement given in Ref. [14] that "the effect of molecular weight on the elution volume is slightly greater in normal-phase than reversed-phase HPLC".)

Figure 6.15 shows the elution and solubility characteristics of S/EMA copolymers in THF/MeOH mixtures. Here, solubility increases with EMA content. The two elution characteristics shown were measured with SEC fractions No. 2 and 5 (see Fig. 6.14) on a C18 column. Fraction No. 5 had lower MW and yielded the dashed curve.

The comparison of Fig. 6.15 with Fig. 6.12 shows that the slope of both solubility and elution characteristics is inverted on changing the phase system. The parallelism between elution characteristic and solubility line should be noticed, as well as the steeper slope of the RP system (indicating higher selectivity in separation by composition).

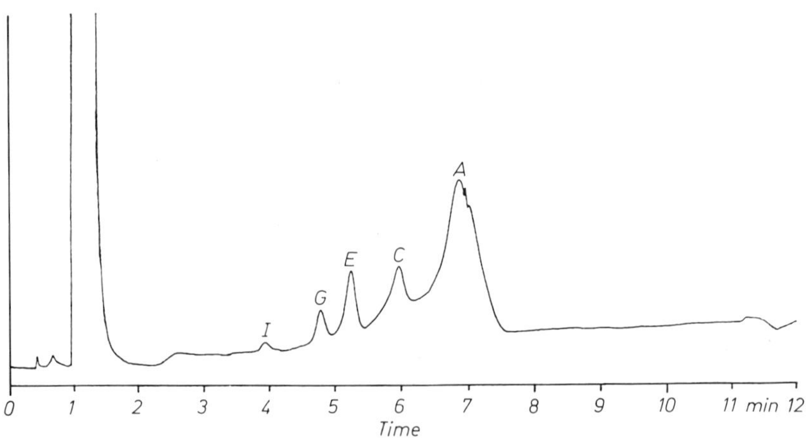

Fig. 6.13. Reversed-phase separation of the mixture of five *stat*-copoly (styrene/ethyl methacrylate) specimens on RP C18 through methanol/tetrahydrofuran gradient (0–100% THF in 10 min) [13]. Column 60 × 4 mm; $d_O \leqq 5$ nm, $d_P = 5\ \mu$m; 50 °C, flow rate 0.5 ml/min; sample: 1.8 μg copolymer A + 1.2 μg C + 2.0 μg E + 1.2 μg G + 2.0 μg I (copolymer data see Table 9.5), in 20 μl THF. UV signal 259 nm, 50 mAU full scale

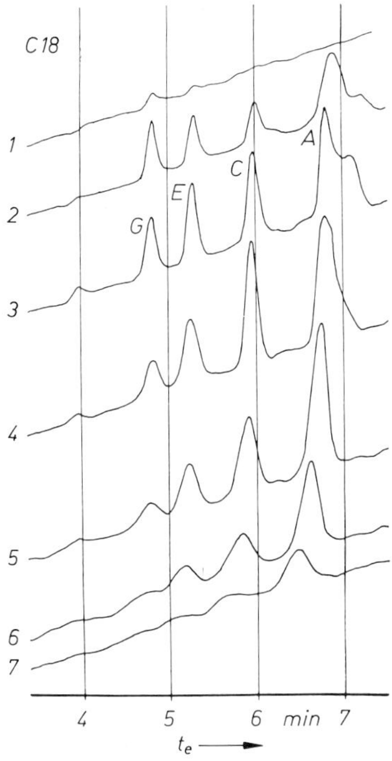

Fig. 6.14. Reversed-phase separation of the mixture of five *stat*-copoly(styrene/ethyl methacrylate) specimens after prefractionation by SEC [19]. Column, flow rate, temperature, gradient, and signal as in Fig. 6.13, starting sample mixture and HPLC injection volume as in Fig. 6.7. Chromatograms redrawn from plots with 25 mAU full scale

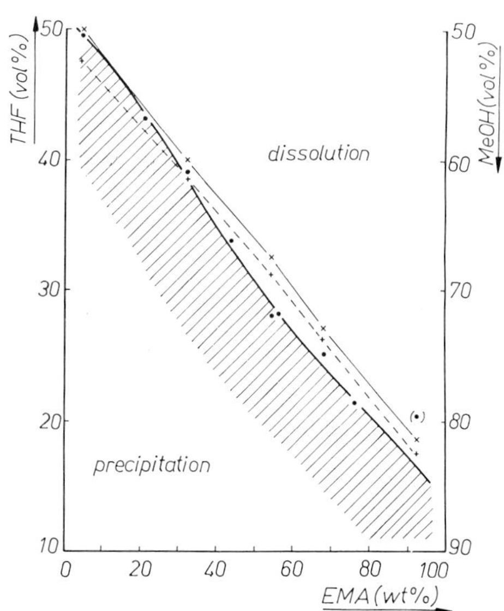

Fig. 6.15. Elution characteristics in methanol/tetrahydrofuran gradients and solubility of *stat*-copoly(styrene/ethyl methacrylate) samples. x: Elution characteristics of fraction No. 2 in Fig. 6.14, +: ditto, fraction No. 5. Solubility borderline measured by turbidimetric titration with methanol at 20 °C

In the NP system shown in Fig. 6.12, the elution characteristics are well above the solubility line. This indicates an adsorption contribution to retention. Sato et al. [15] investigated S/MMA samples by gradient elution on different columns and have also recognized an adsorption mechanism on polar packings.

In contrast to the shift visible in Fig. 6.12, RP elution occurs almost precisely at the solubility borderline. Retention in RP chromatography is often referred to as solvophobic retention [16]. In a polymeric systems precipitation is likewise a solvophobic phenomenon.

Teramachi et al. [17] succeeded in NP and RP separation of S/MMA copolymers and also found inversion of elution order: the sample with highest content in MMA was retained longest in NP mode but eluted first in a RP system, see Figs. 9.8 and 9.9. With S/MMA copolymers in AcN/DCM gradients on cross-linked PS, Sato et al. [14] also observed inversion of elution order in RP HPLC of S/MMA as compared with previous work in NP systems [15].

6.4 References

1. Larmann JP, DeStefano JJ, Goldberg AP, Stout RW, Snyder LR, Stadalius MA (1983) J Chromatogr 255: 163
2. Tennikov MB, Nefedov PP, Lagareva MA, Frenkel SJa (1977) Vysokomol Soed Ser A19: 657
3. Entelis SG, Evreinov VV, Gorshkov AV (1986) Adv Polym Sci 76: 129
4. Mourey TH (1983) (private communication)
5. Klein J, Leidigkeit G (1979) Makromol Chem 180: 2753
6. Lai S-T, Sangermano L, Chou CK, Locke DC (1985) J Chromatogr 324: 436
7. Balke ST, Patel RD (1983) In: Craver CD (ed) ACS Advances in Chemistry Series 203, Amer Chem Soc
8. Gorshkov AV, Evreinov VV, Lausecker B, Pasch H, Becker H, Wagner G (1986) Acta Polym 37: 740
9. Evreinov VV, Gorshkov AV, Prudskova TN Gur'yanova V, Pavlov AV, Malkin AYa, Entelis SG (1985) Polym Bulletin 14: 131
10. Sato H, Takeuchi H, Suzuki S, Tanaka Y (1984) Makromol Chem, Rapid Commun 5: 719
11. Glöckner G, van den Berg JHM (1987) J Chromatogr 384: 135
12. Glöckner G, van den Berg JHM (1986) J Chromatogr 352: 511
13. Glöckner G (1987) J Chromatogr 403: 280
14. Sato H, Mitsutani K, Shimizu I, Tanaka Y (1988) J Chromatogr 447: 387
15. Sato H, Takeuchi H, Tanaka Y (1986) Macromolecules 19: 2613
16. Horváth Cs, Melander W, Molnar I (1976) J Chromatogr 125: 129
17. Teramachi S, Hasegawa A, Motoyama K (1987) Polym Prepr Japan (Engl Ed) 36: E441, 3169
18. Glöckner G, Stickler M, Wunderlich W (1987) Fresenius Z Anal Chem 328: 76
19. Glöckner G, Stickler M, Wunderlich W (1988) Fresenius Z Anal Chem 330: 46

7 Detection in Gradient High-Performance Liquid Chromatography

In gradient elution, detection requires quantitative measurement of the sample components in an eluent whose composition and, hence, physical properties alter in the course of the analysis. Even solvent combinations purposefully selected with respect to a certain property, e.g. "iso-refractive" solvents, cannot ensure proper measurements of small solute concentrations by detectors which monitore a bulk property of the mobile phase plus the solute.

The detection problem in gradient elution can be solved by either using a selective detector sensitive to a property of only the solute (see Sect. 7.1) or stripping off the eluent with subsequent measurement of non-volatile residues (see Sect. 7.2).

The current status and further prospects of HPLC detectors have been dealt with in a recent review article [1], which compares the limits of detection for twelve different kinds of detectors, and in monographs [2, 3].

7.1 Selective Detectors

Selective devices are (1) UV/visible photometers and spectrophotometers, (2) infrared photometers, (3) amperometric detectors, (4) reaction detectors, as well as radioactivity detectors, fluorometers, optical rotation detectors, circular dichroism detectors, and laser light scattering detectors.

7.1.1 UV/Visible Photometers and Spectrophotometers

Photometers for UV and visible light are most common detectors in HPLC. The cells typically have an optical path length of 10 mm and a volume of about 8 μl. The flow path is Z-shaped in order to minimize stagnant regions in the cell. Quartz windows are pressed against the core of the cell.

The signal from cylindrical cells may be influenced by changes in refractive index which form "liquid lenses" deflecting light towards the cell walls. They reduce the energy reaching the light sensor and produce pseudopeaks which do not belong to UV absorption [4]. The effect is especially pronounced at sudden changes in gradient steepness. Modern absorption detectors have tapered flow cells with diverging walls [5] where light is prevented from hitting the walls even if the refractive index of the mobile phase is changing.

The windows of a flow cell may sometimes be not strictly parallel. Wedge angles varying from 0.5 to 2° were found in a painstaking investigation [6]. A wedged cell behaves like a prism and, thus, also causes light deviation on changing refractive index. The transmitted beam wanders around on the surface of the photodiode where local response inhomogeneities can generate small changes in signal size due to changes of the mobile phase or to incomplete mixing in the detector cell.

7.1.2 Infrared Photometers

Detectors of this kind permit stable operation with a heated cell up to 150 °C. They are suited for high-temperature separations of synthetic polymers and are used, e.g. in SEC of polyolefins or temperature-rising elution fractionation (see Sect. 9.14).

In gradient HPLC, IR detectors can be used when eluent mixtures are available which have suitable "windows", i.e. no absorption in a wavenumber range where the copolymer can be detected. Simple copolymers of known structure can be analysed this way if suitable eluents can be found.

For instance, an IR detector has been used for selectively monitoring at $1730\,cm^{-1}$ the MMA content of S/MMA copolymers in gradient elution with hexane/chloroform by measuring the carbonyl absorption. The styrene content could be monitored by UV absorption at 254 nm, see Fig. 9.4 [7].

In general, the gradient components may disturb the application of IR detection. In that case a deposition technique can be applied which also enables the whole IR spectra of the eluite to be measured. In this technique, the column eluate is sprayed onto a substrate, which is slowly moved beneath the effluent spray. The solvents are vaporized by heating. Nonvolatile sample components remain spatially separated on the substrate and can be analysed by Fourier transform IR. The technique has been used so far with supercritical fluid chromatography [8–11].

7.1.3 Amperometric Detectors

A device of that kind requires conductivity of the eluent. This problem can be circumvented by postcolumn addition of a solvent with a high dielectric constant plus supporting electrolyte.

Nitrocellose could be selectively detected by electrochemical reduction at a pendent mercury drop electrode held at zero potential vs Ag/AgCl electrode. An appropriate level of insulation at a minimal contact with oxygen-permeable PTFE allowed for a background current not larger than 2×10^{-9} A. The device enabled nanogram amounts of nitrocellulose to be characterized [12].

7.1.4 Reaction Detectors

Chemical derivatization by post-column reactions is another approach to selective detection. For a review, see References [13, 14] and the literature cited there.

With synthetic polymers, post-column reaction-detection has been performed with the help of an ozone reaction detector which monitores double bonds. The device has been employed in SEC investigations of polybutadiene and EPDM rubber [15] as well as of polyisobutylene [16].

Post-column precipitation of the solute provides another possibility of selective detection. A nephelometric HPLC detector has been described which uses a helium-neon laser and measures, under an angle of 90° and in a 17 μl flow-through cell, the light scattered by the effluent after addition of a precipitant. The instrument was used for the detection of non-polar lipids by continuously mixing the acetone + MeOH (2:1) effluent with an aqueous solution of ammonium sulfate [17]. In a similar manner, proteins have been monitored [18]. The principle looks promising for polymers since they have exceptionally narrow solubility limits. The necessity of gradient elution may render a realization more difficult than under isocratic conditions.

7.2 Detection Subsequent to Eluent Vaporization

7.2.1 Transport Detectors

The sensitivity and selectivity of several detectors used in gas chromatography were the reasons for employing them in liquid chromatography, too. For this purpose, the eluent has to be removed from the solute. This can be achieved by evaporation if the boiling points of solute and mobile phase are sufficiently different. With suitable solvents, this holds true for lipids, fatty acids, saccharides, higher alcohols, polycyclic hydrocarbons and polymers (which are not volatile at all).

The effluent is either coated onto a chain [19], a moving wire [20–22], an endless belt of stainless steel [23], a spiral [24], or soaked into a quartz belt mounted on a rotating disk [25]. After evaporation of the solvent, the remainders are transported to a pyrolyser.

In a formerly commercial moving-wire detector [26], a stainless-steel wire (0.122 mm in diameter) passed at a speed of typically 10 cm/s through a cleaner oven, a coating block (where about 0.2 μl eluate per second were loaded onto the wire), an evaporator oven ($\leq$ 300 °C) and a pyrolyser oven whose temperature was, according to the problem to be solved, set to a value in the range 300–750 °C. With about 10 km of wire, each reel permitted about 30 h net registration time.

Pyrolysis in an inert atmosphere as well as burning off the sample were used for transforming non-volatile solutes into gaseous products. In an improved version of the moving-wire detector [27], the carbon content of a sample was oxidized to carbon dioxide which was subsequently hydrogenated to methane in the presence of a nickel catalyst.

The moving-wire detector was preferably used in connection with a flame-ionization detector (FID) but combinations with an argon ionization detector or, when selectivity to halogens was desired, with an electron-capture detector, were also commercially available [28].

The drawback of moving-wire detectors was the limited amount of effluent that could be loaded. By spraying the effluent onto the wire instead of coating, the sensitivity of a moving-wire/FID could be increased by a factor of 20–50 [29]. Nebulization has been also suggested with almost complete vaporization of the solvent before deposition onto a moving belt of stainless steel [30]. On the other

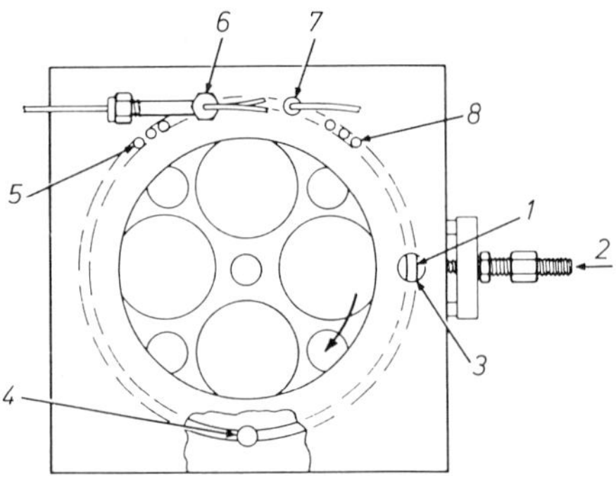

Fig. 7.1. Schematic diagram of the 945 Universal FID Detector for HPLC. *1*: Quartz belt, *2*: column effluent, *3*: application, *4*: vacuum port, *5*: evaporation air in, *6*: detector flame and FID; *7*: cleaning flame; *8*: cooling air in. From Ref. [25] with permission

hand, the small effluent volume required made the moving-wire principle attractive for detection in microbore HPLC with flow rates of 60–100 μl/min [31].

The "945 LC/FID" detector [25] contains a slowly rotating metal disk (approximately 15 cm in diameter) with a fibrous quartz belt on its circumference, see Fig. 7.1. The disk is enclosed in a heated, air-swept housing. At each revolution, the quartz belt is cleaned by a high-temperature flame. After cooling the belt with air, the column effluent is applied into the quartz fiber belt. The disk rotates further within the housing to the zone where the volatile solvents are evaporated. Non-volatile solutes remain in the belt and are detected as the disk rotates further past the dual flame ionisation detectors. Remaining residues are burned off by a much hotter oxidative cleaning-flame as the disk continues to rotate. A sensitivity of 1 μg is reported for organic compounds. The quartz-belt detector proved successful in the investigation of ethylene oxide oligomers [32] and in the gradient HPLC of lipids [33–35]. A linear response over a range 6 to 200 μg has been reported [35].

7.2.2 Evaporative Light-Scattering Detectors

An evaporative light-scattering (ELS) detector [36–39] enables the quantification of any non-volatile solute in every eluent which can be vaporized under operating conditions. The eluate from the HPLC column is directed into the nebulizer where a gas supply atomizes the liquid stream. The droplets are sprayed into the evaporator where heat vaporizes the solvent. Solutes less volatile than the solvent remain as a cloud of fine particles which are carried down the evaporator past the light scattering device. Here, collimated light is passed through the instrument at right angles to the direction of gas flow. The light scattered from the dust particles is measured by a photomultiplier.

An ELS detector designed by Stolyhwo et al. [39] is equipped with a helium-neon laser and the scattered light is collected by a glass rod which is placed perpendicular to the laser beam at a distance of 2–5 mm from the scattering volume. The tip of the rod (diameter approx. 5 mm) has a cylindrical, concave shape and is mirror polished [40].

A polarized laser beam is used in an improved version of the detector. Horizontal orientation of the polarization plane yielded the lowest response but the widest dynamic range [40]. Towards the laser, the detector is tightened by a glass window in order to exclude the dust in the laboratory atmosphere from the optical system. Figure 7.2 provides a schematic diagram of the ELS detector. The detection limit was 4.5 ppm [41] or even 1 ppm [40]. (The value given first corresponds to 3 ng). The temperature of the drift tube had no effect on the response provided that the heat supply was sufficient for complete evaporation of the solvent without vaporizing the sample [42].

In a commercial ELS detector [43], the light source is a tungsten filament lamp and a photomultiplier measures the scattered light around an angle of 60°. (The scattering angle defines the orientation of the light sensor with respect to the direction of the primary beam. The location of a photomultiplier "at an angle of 120° to the incident light beam" [43, 44] implies a scattering angle of 60°.) A scheme of the commercial instrument is shown in Fig. 7.3.

Nitrogen or air can be used as nebulizing gases. The former is recommended with inflammable mobile phases. Stolyhwo et al. [39] used carbon dioxide and directed the exhaust from the detector to a water aspirator in order to avoid any harm to the environment.

The two main stages in the operation of an ELS detector are (i) the transformation of the column effluent in an aerosol and (ii) and measurement of any non-volatile residues of the droplets.

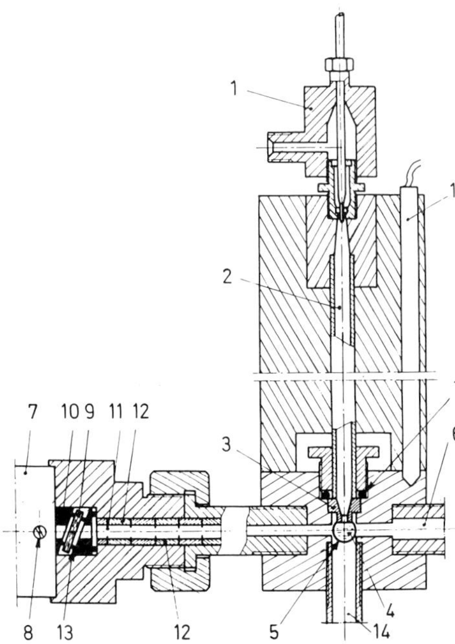

Fig. 7.2. Schematic diagram of a laser-beam evaporative light-scattering detector. *1*: nebulizer; *2*: drift tube, 300 × 2 mm I.D.; *3*: nozzle for focussing the droplets to the laser beam; *4*: glass-rod light-collector; *5*: opaque coating; *6*: outlet to light trap; *7*: helium-neon laser; *8*: shutter; *9*: glass window; *10*: mounting of the glass window; 11: apertures, to eliminate divergent non-coherent light; *12*: spacers; *13*: sealing O-rings, *14*: exhaust for the driving gas and suspended particles; 15: heating cartridge; *16*: sealing O-ring. From Ref. [40] with permission

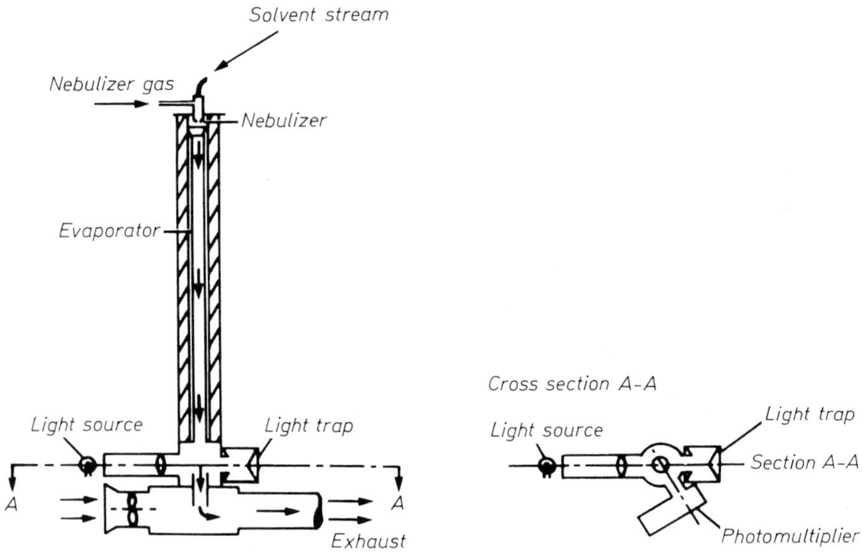

Fig. 7.3. Schematic diagram of a commercial evaporative light scattering detector (ACS Model 750/14) with indication of the position of the photomultiplier. From Ref. [43] with permission

The aerosol is produced by a Venturi nebulizer. The test of different devices yielded favorable results with a concentric design as shown in Fig. 7.2. The nebulizer had a strong effect on detector response. Replacing the nebulizer in a given detector by another one of the same design caused, on injection of 10 μg olive oil in acetone, the response to vary in a 1:4 range. The response was stable with both nebulizers even on a long-term basis but recalibration was necessary when the position of the liquid nozzle was changed, e.g. after cleaning and reassembling the device [42].

The droplet size is mainly governed by the linear velocity of the nebulizing gas in the channel surrounding the effluent stream. Increasing gas velocity decreases the average diameter of the aerosol and thus the detector response [45, 46]. Subsonic velocities in the Venturi tube produced baseline noise, presumably because of the formation of too large droplets which vaporized incompletely [42, 45]. Stable results were obtained with supersonic velocities [39, 42, 45].

Mourey and Oppenheimer modelled the response of ELS detectors by applying nebulizing and light scattering theories [45, 47]. The distribution of particle size in an aerosol produced by a Venturi nebulizer can be described through an upper-limit log-normal distribution [48]:

$$\frac{\mathrm{d}N}{\mathrm{d}d} = \frac{\mathrm{C}_1 d_\mathrm{m}}{\sqrt{\pi} d(d_\mathrm{m} - d)} \exp\left[\mathrm{C}_1 \ln \frac{\mathrm{C}_2 d}{d_\mathrm{m} - d} + \frac{3}{2\mathrm{C}_1}\right]^2 \tag{7.1}$$

Here, N is the number of droplets with diameter d, C_1 and C_2 are parameters of the distribution and d_m is the diameter of the largest drops formed by the nebulizer, which is related to the Sauter mean diameter, d_s, by

$$d_m = d_s\left[1 + C_2 \exp\frac{1}{4C_1^2}\right] \tag{7.2}$$

For a commercial ELS detector and MEK as a solvent, Mourey and Oppenheimer [45] found an estimate of 0.6 for both C_1 and C_2. Other liquids may form different solvent distributions; thus, the values of C_1 and C_2 need not be universal.

The Sauter mean diameter can be approximated [49–51] by:

$$d_s = \frac{585\sqrt{\sigma}}{(u_g - u_l)\sqrt{\rho}} + 597\left(\frac{\eta}{100\sqrt{\sigma\rho}}\right)^{0.45}\left(\frac{1000F_l}{F_g}\right)^{1.5} \tag{7.3}$$

where u_g and u_l are the linear velocity (in m/s) of the gas and liquid stream in the Venturi nebulizer, σ the surface tension of the liquid in mN/m (or dyn/cm), ρ its density in g/cm^3, η its viscosity in mPas (or cPoise), and F_l/F_g the ratio of liquid to gas volumetric flow rate. Equation (7.3) can be applied when the surface tension of the liquid is in the range 19 to 73 mN/m, its density between 0.7 and 1.2 g/cm^3, and its viscosity in the range 0.3 to 50 mPas.

A droplet of diameter d forms, after evaporation of the liquid, a particle of diameter d_P:

$$d_P = d\left(\frac{c_2}{1000\rho_2}\right)^{1/3} \tag{7.4}$$

Here, c_2 is the solute concentration in the effluent (in g/l) and ρ_2 the solute density in g/cm^3. Increasing solute concentration increases the size of the particles rather than their total number. With a reasonable estimate of solute concentration ($10^{-2} \leqq c_2 \leqq 1$ g/l) and density, the diameter of the remaining particles will be about 2 – 10% of the initial droplets.

Oppenheimer and Mourey [47] calculated the detector response as a function of solute concentration on base of Mie's scattering theory, i.e. for a particle-to-wavelength ratio between 0.1 and 10. The result, normalized to a constant volume of nebulized effluent, are plotted as log(signal) vs log(concentration) in Fig. 7.4.

The slope of the curves decreases with solute concentration. At the lowest concentration taken into account, it is about 2 for $d_s = 5\,\mu$m or 1.87 for $d_s = 20\,\mu$m, and in the high-concentration region 0.94 or 0.61, respectively. The calculated slope data approximately match experimental values: 1.81 [39], 1.69 [40], or unity [37, 45, 52].

A curved logarithmic response characteristic was also found experimentally. In Fig. 7.4, data measured by Stolyhwo et al. [40] or Schultz [46] have been added to the theoretical curves by a fitting procedure which assumes proportionality of the response units used as well as of concentration units. Although the commercial instrument used in Ref. [46] measured under an angle of 60°, the correspondence with data measured [40] or calculated [47] for 90° scattering angle is sufficient.

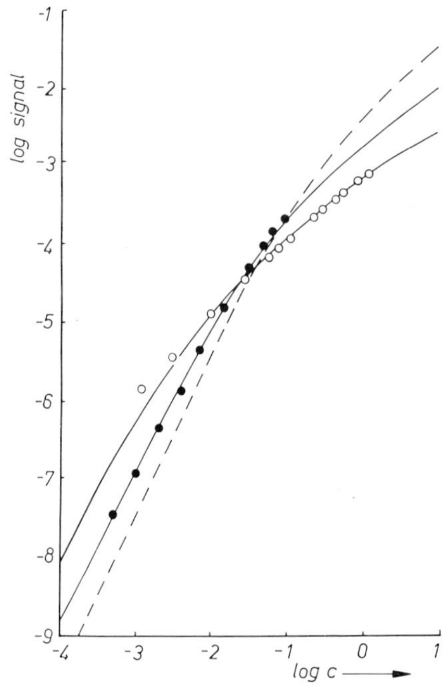

Fig. 7.4. Response curves for detection at 90° using unpolarized light (632.8 nm wavelength), calculated from the Mie theory and the particle distribution given by Eqs. (7.1) and (7.4). Sample refractive index: 1.57, density: 1.05 g/cm^3 (data of polystyrene). The Sauter mean diameter of the droplets was assumed to be $d_s = 5\,\mu m$ (*dashed line*), 10 μm, or 20 μm [47]. The *points* result from measurements of dioctylphthalat in MeOH (•, [40]) or PS in DCM (o, [46]) and have been positioned to the indicated scale by appropriate shift of their abscissa and ordinates

Stolyhwo et al. [40] found a logarithmic response characteristic with a constant slope of 1.69 over a 500:1 range of sample size. Within this range, quantitative evaluation of chromatograms could be performed by using $m = C_1 A^{1/1.69}$ where A is the peak area and m the sample size.

Over about a decade in concentration, the slope is close to unity (see Fig. 7.4), but a linear dependence of detector response on sample concentration in the overall range is neither predicted by theory nor found experimentally. Thus (and also because of a solvent effect), the ELS detector is not a mass detector [42, 46] and care must be taken with signals at the bonds of the interval used.

At low concentrations, the signal and thus the sensitivity of the detector increases with droplet size. At high concentrations, the calculated curves indicate a larger signal at reduced droplet size. Thus, in the entire range of concentration, the sensitivity of the instrument cannot be increased by increasing the particle size.

With a carefully adjusted instrument, the main source of baseline noise was the presence of non-volatile impurities in the solvents [40].

ELS detectors have been successfully employed for monitoring the HPLC separation of lipids [41, 53–57], carbohydrates [38, 58, 59], coal derivatives [60], hydrocarbons [61] and petroleum fractions [62]. They have been used for detecting polymers in SEC [63, 64] and, above all, in gradient HPLC [46, 65, 66]. ELS detection in sedimentation field flow fractionation of latex particles has also been reported [67]. Here, the detector proved advantageous with samples containing particles less than 0.2 μm in diameter. Another interesting application is the use of an ELS detector for control of silica-based bonded-phase columns under the influence of W/AcN gradients [68].

A demonstration of the advantage of ELS detection is given in Figs. 7.5 and 7.6 which show the RP separation of butter triglycerides by a gradient EtOH/AcN [53]. The chromatogram in Fig. 7.5 was monitored by UV detection at 225 nm. Even at this short wavelength the sample components yielded only small peaks while the baseline showed a steep slope due to the gradient. Figure 7.6

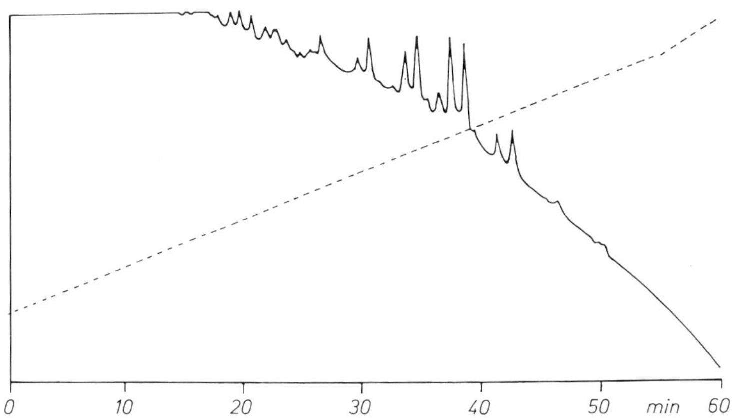

Fig. 7.5. Gradient HPLC of butter triglycerides monitored by UV signal at 225 nm. Chromatographic conditions: a set of two C18 columns (Spherisorb-5 ODS2, 50 × 4.6 and 250 × 4.6 mm), gradient ethanol/acetonitrile (20–100% in 60 min, *dashed line*), flow rate 1.5 ml/min. From Ref. [53] with permission

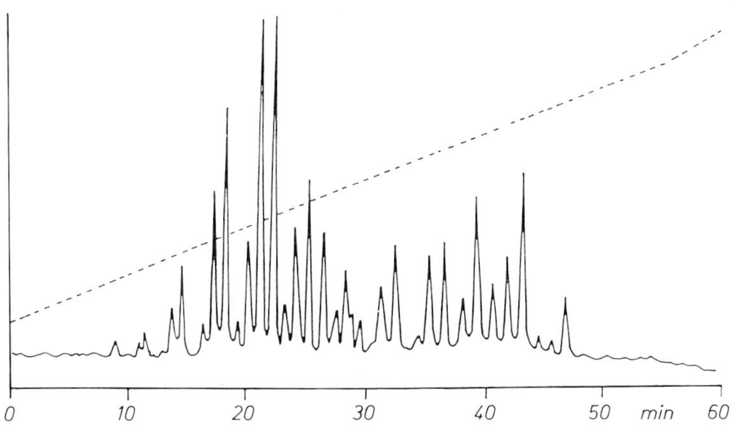

Fig. 7.6. Same separation as shown in Fig. 7.5, but monitored by evaporative light-scattering detector 750/14. From Ref. [53] with permission

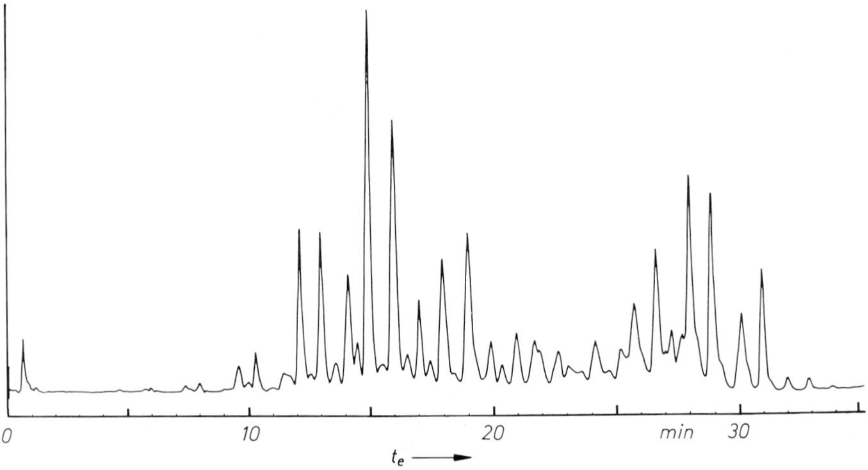

Fig. 7.7. Separation of butter triglycerides by reversed-phase gradient elution (acetonitrile/acetone, 33–99% in 25 min). column 200 × 2 mm ID, packed with LiChrosorb RP18, $d_P = 5$ μm, flow rate 0.3 ml/min. Chromatogram monitored by laserbeam evaporative light-scattering detector. From Ref. [39] with permission

was obtained under identical chromatographic conditions but with a commercial ELS detector which yielded a stable baseline and good response.

Figure 7.7 shows the RP separation of a different sample of butter triglycerides by a gradient AcN/acetone. In this investigation, a laser-beam ELS detector was employed [39]. In spite of the differences in sample origin as well as in chromatographic conditions, the results obtained with two different version of ELS detectors compare favourably. (With modified chromatographic conditions and an improved version of an ELS detector a still higher resolution of butter triglycerides was obtained [40].)

With non-volatile solutes (boiling points above 280–300 °C) the detection level of the ELS detector corresponds to that of the refractive index detector [39] or is even more favourable [40].

Changes in eluent composition may, in general, yield different values of effluent surface tension, density, and viscosity and thus, by Eq. (7.3), different values of the Sauter mean diameter. Furthermore, in Eqs (7.1 and 7.2) the values of C_1 and C_2 may change. Mourey and Oppenheimer [45] measured the detector response with polystyrene as a solute in toluene, MEK, THF, or DCM and found good correspondence of results measured with Tol, MEK, and THF (relative standard deviation 5.4% or 0.6% at sample concentration 0.07 g/l or 4 g/l, respectively) but a rather large deviation of the result determined in DCM from the average obtained with the other solvents (−25.0% or −26.2%, respectively). Schultz [46] investigated PS in DCM or MeOH/DCM and Hp/DCM gradients and also observed diverging results in DCM solutions.

The density and refractive index of the sample also effects the detector response. Hence, calibration with appropriate solvents and samples is necessary for quantitative evaluation of chromatograms. With copolymers of different composition a constant mass response may be obtained under favourable conditions. Schultz investigated S/AN copolymers and did not find a systematic composition effect on detector response in the range of 2.3–37 mass% AN [46].

7.3 References

1. Yeung ES (1989) LC–GC International 2: 2/38
2. Scott RPW (1986) Liquid chromatography detectors, 2nd edn, Elsevier, Amsterdam
3. Yeung ES (1986) (ed), Detectors for liquid chromatography, John Wiley, New York
4. Evans CE, Shabushnig JG, McGuffin VL (1988) J Chromatogr 459: 119
5. Little JN, Fallick GJ (1975) J Chromatogr 112: 389
6. Peck K, Morris MD (1988) J Chromatogr 448: 193
7. Sato H, Takeuchi H, Tanaka Y (1986) Macromolecules 19: 2613
8. Shafer KH, Pentoney SL, Griffiths PR (1984) J High Resol Chromatogr, Chromatogr Commun 7: 707
9. Fujimoto C, Hirata YH, Jinno K (1985) J Chromatogr 332: 47
10. Pentoney SL, Shafer KH, Griffiths PR (1986) J Chromatogr Sci 24: 230
11. Shafer KH, Griffiths PR, Pentoney SL, Fuoco R (1986) J High Resol Chromatogr, Chromatogr Commun 9: 168
12. Lloyd JBF (1986) J Chromatogr 351: 323
13. Frei RW, Lawrence JF (1981/82) (eds) Chemical derivatization in analytical chemistry, Plenum, New York
14. Brinkman UA Th, Frei RW, Lingeman H (1989) J Chromatogr 492: 251
15. Poznjak TI, Lisicyn D, Novikov DD, D'jackovskij FS (1977) Vysokomolekularnye Soedinenja, Ser A, 19: 1168
16. Poznjak TI, Lisicyn D, Novikov DD, Berlin AA, D'jackovskij FS, Procuchan JuA, Sangalov, JuA, Minsker KS (1980) Vysokomolekularnye Soedinenja, Ser A 22: 1424
17. Jorgenson JW, Smith SL, Novotny M (1977) J Chromatogr 142: 233
18. Tappan DV (1966) Anal Biochem 14: 171
19. Haahti EOA, Nikkari T (1963) Acta Chem Scand 17: 2565
20. James AT, Ravenhill JR, Scott RPW (1964) Chem Ind 18: 746
21. Karmen A (1966) Anal Chem 38: 286
22. Maggs RJ (1968) Chromatographia 1: 43
23. Privett OS, Erdahl WL (1978) Anal Biochem 84: 449
24. Stolyhwo A, Privett OS, Erdahl WL (1973) J Chromatogr Sci 11: 263
25. Tracor 945 LC/FID, Tracor Instruments, Austin (Texas), USA US Patents # 4,271,022 and 4,215,090
26. Pye LCM2 Liquid Chromatograph, PYE UNICAM Ltd, Cambridge, UK, (1661/7.5 m/10.71)
27. Scott RPW, Lawrence JG (1970) J Chromatogr Sci 8: 65
28. Pye Liquid Chromatograph Systems, Cat No 14001,002,003, PYE UNICAM Ltd, Cambridge, UK, (83/1.3½ M/9.69)
29. van Dijk JH (1972) J Chromatogr Sci 10: 31
30. Yang L, Fergusson GJ, Vestal ML (1984) Anal Chem 56: 2632
31. Veening H, Tock PPH, Kraak JC, Poppe H (1986) J Chromatogr 352: 345
32. McClure JD (1982) J Amer Oil Chemists' Soc 59: 364
33. Smith LA, Norman HA, Cho SH, Thompson GA (1985) J Chromatogr 346: 291
34. Norman HA, St John JB (1986) J Lipid Res 27: 1104
35. Maxwell RJ, Nungesser EH, Marmer WN, Foglia TA (1987) LC–GC International 1: 56
36. Ford DL, Kennard W (1966) J Oil Colour Chem Assoc 49: 299
37. Charlesworth JM (1978) Anal Chem 50: 1414
38. Macrae R, Dick J (1981) J Chromatogr 210: 138
39. Stolyhwo A, Colin H, Guiochon G (1983) J Chromatogr 265: 1
40. Stolyhwo A, Colin H, Martin M, Guiochon G (1984) J Chromatogr 288: 253
41. Stolyhwo A, Martin M, Guiochon G (1987) J Liquid Chromatogr 10: 1237
42. Guiochon G, Moysan A, Holley Ch (1988) J Liquid Chromatogr 11: 2547

43. Model 750/14, Applied Chromatography Systems Limited, Macclesfield, Cheshire, UK; Zinsser Analytic, Frankfurt/Main, BRD; Peris Industries, State College, Pennsylvania, USA
44. Turner B (1986) Laboratory Practice 55: 35
45. Mourey TH, Oppenheimer LE (1984) Anal Chem 56: 2427
46. Schultz R (1989) Thesis, University of Saarbrücken
47. Oppenheimer LE, Mourey TH (1985) J Chromatogr 323: 297
48. Bitron MD (1955) Ind Eng Chem 47: 23
49. Nukiyama S, Tanasawa Y (1938) Trans Soc Mech Eng Japan 4: 86, 138
50. Nukiyama S, Tanasawa Y (1939) Trans Soc Mech Eng Japan 5: 63, 68
51. Nukiyama S, Tanasawa Y (1940) Trans Soc Mech Eng Japan 6: 117
52. Bear GR (1988) J Chromatogr 459: 91
53. Robinson JL, Macrae R (1984) J Chromatogr 303: 386
54. Christie WW (1986) J Chromatogr 361: 396
55. Sortirhos N, Thorngen C, Herslof B (1985) J Chromatogr 331: 313
56. Robinson JL, Tsimidou M, Macrae R (1985) J Chromatogr 324: 35
57. Grossberger T, Rothschild E (1989) LC–GC International 2/7 (1989) 44 and LC–GC 7: 439
58. Burns ID, Jones DA Unilever Research, Sharnbrook, UK, (Appl Chrom Systems, Application Bulletin No 60)
59. Macrae R, Trugo LC, Dick J (1982) Chromatographia 15: 476
60. Bartle KD, Mulligan MJ, Taylor N, Martin TG, Snape CE (1984) Fuel 63: 1556
61. Lafosse M, Dreux M, Morin-Allory L (1987) J Chromatogr 404: 95
62. Coulombe S (1988) J Chromatogr Sci 26: 1
63. Smith BR (1976) Rubber Chem Technol 49: 278
64. Huang SS, Barth HG (1985) Soc Plastics Engineers, Annual Techn Conf, Techn Papers 277
65. Mourey TH (1986) J Chromatogr 357: 101
66. Augenstein M, Stickler M (1990) Makromol Chem 191: 415
67. Oppenheimer LE, Mourey TH (1984) J Chromatogr 298: 217
68. Lafosse M, Herbreteau B, Dreux M, Morin-Allory L (1989) J Chromatogr 472: 209

8 Quantitative Evaluation and Calibration

There are several pitfalls in polymer HPLC which can cause unnoticed loss of sample, incomplete elution, or imperfect separations. Effects of that kind not only make quantitative evaluation illusive but also may severely falsify even qualitative conclusions. Since these effects disturb the common linear dependence of peak area on sample size, quantification of suitable calibrating chromatograms is strongly recommended.

8.1 Features of a Proper Run in Polymer HPLC

As already mentioned in the discussion of Fig. 1.6, the first peak in UV recordings of polymer gradient HPLC is usually a solvent peak. Even when the sample solution can be prepared in a portion of the starting eluent, a UV detector often responds to the solvent plug due to slightly differing properties. With UV detectors sensitive to changes in refractive index, the signal of the solvent plug is usually falsified, possibly even with a negative deflexion of the signal. Only with detectors equipped with a tapered cell (see Sect. 7.1.1), quantitative evaluation of UV solvent peaks is feasible.

Figure 8.1 shows the traces of S/EMA copolymers on a CN bonded-phase column in gradient *i*Oct/THF. The peaks at 1.2 min elution time are solvent peaks. The solvent plug is generally eluted at a V_e value corresponding to the mobile-phase volume in the column. The gradient appears later although it is usually started simultaneously with the injection. The delay is due to the difference between *dead time* and gradient *dwell time*. In a proper run, a polymer sample should completely elute during the gradient period, not too close to the beginning or the end of the latter.

8.2 Quantitative Evaluation

The quantitative evaluation of low-MW HPLC is facilitated by high and narrow peaks. Quantification of polymer HPLC is usually more difficult because, due to the distribution in size and composition of the samples, broad peaks must be evaluated. Electronic integration is reliable only if the position of the baseline can be controlled, which is possible with modern HPLC workstations. Without baseline control, automatic integration can yield misleading results. Special care

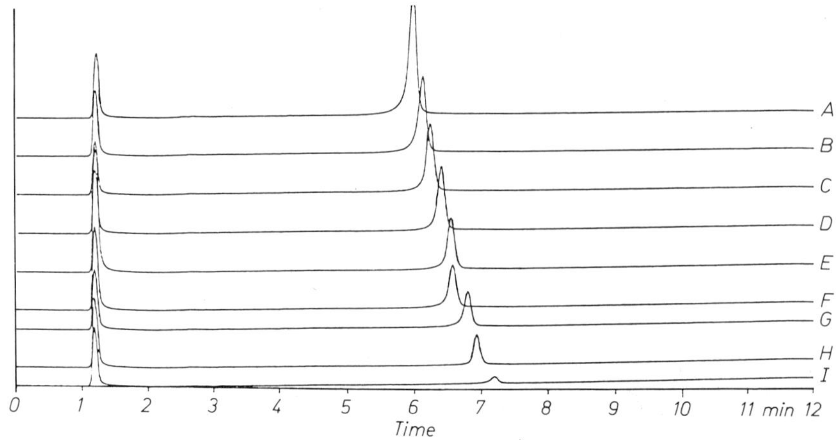

Fig. 8.1. Gradient elution of *stat*-copoly(styrene/ethyl methacrylate) samples on a CN bonded-phase column (60 × 4 mm; $d_O \leqq 5$ nm; $d_P = 5$ μm) after separate injections of 10 μg each in THF, $V_O = 10$ μl. Sample specification see Table 9.5; gradient: *iso*-octane/tetrahydrofuran (0–100% in 10 min), flow rate 0.5 ml/min, 50 °C, UV signal 259 nm. Chromatograms plotted with 500 mAU full scale. By courtesy of Friedr. Vieweg & Sohn Verlagsgesellschaft [2]

is required with nonlinear baselines due to, e.g. sophisticated gradient programs. Here, point-by-point reading of the difference h_i between chromatogram and blank-gradient baseline is necessary. When in doubt, this cumbersome procedure must eventually be performed manually. Totalling yields peak area

$$A = \sum h_i \Delta t \tag{8.1}$$

and retention time (first moment):

$$t_e = \frac{\sum h_i t_{e,i}}{\sum h_i} \tag{8.2}$$

With UV detection, the measured peak area is in mAUs and can eventually be compared with the absorption calculated from the molar absorptivity ε. Lambert–Beer's equation reads

$$E = \log \frac{I_0}{I} = \varepsilon \cdot l \cdot c \tag{8.3}$$

where I_0 and I are the intensity of incident and transmitted light, respectively, l the cell path length (in cm) and c the concentration of absorbing solute (in mol/l).

The absorbance unit (AU) is that value of the product $\varepsilon l c$ which causes I_0 to decrease to $I = I_0/10$.

By multiplying with Ft, Eq. (8.3) yields

$$E \cdot F \cdot t = \varepsilon \cdot l \cdot c \cdot F \cdot t = \varepsilon \cdot l \cdot m_0 \tag{8.4}$$

where F is the flow rate, t the length of period that covers the elution of a peak, Et the peak area in mAUs and m_0 the mass injected.

In Eq. (8.4), Et and ct represent the total of infinitesimal small contributions $E(t)\Delta t$ and $c(t)\Delta t$. At constant flow rate and with two solutes, we have

$$(E\cdot t)_1/(E\cdot t)_2 = (\varepsilon_1\cdot m_{0,1})/(\varepsilon_2\cdot m_{0,2}) \tag{8.5}$$

or, since $A = Et$,

$$A_1 = \varepsilon_1\cdot m_{0,1}\left[\frac{A_2}{\varepsilon_2\cdot m_{0,2}}\right] \tag{8.6}$$

The term in square brackets is determined by the geometry of the cell and can be calculated, in principle, from path length and cell volume, but measuring the value $A_2/(\varepsilon_2 m_{0,2})$ by using a substance of known absorptivity is more convenient and even more reliable because, in a real detector, the measured ratio I_0/I is to some degree influenced by stray-light and dark-current contributions [1].

8.3 Incomplete Retention and Elution

Incomplete retention of a sample polymer can cause (i) excluded elution, (ii) elution in a peak which immediately follows the solvent peak, or (iii) elution together with the sample solvent. Figure 8.2 shows the UV recording after injection of 100 μl of a SEC fraction containing 24.4 μg of the mixture of five

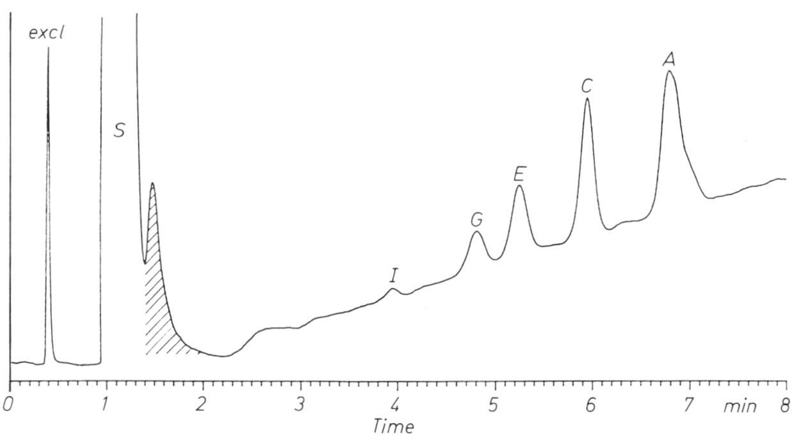

Fig. 8.2. Incomplete retention of *stat*-copoly(styrene/ethyl methacrylate) specimens on injection of $V_0 = 100\,\mu$l sample solution into methanol on a C18 bonded-phase column (60 × 4 mm; $d_0 \leqq 5$ nm; $d_P = 5\mu$m). Sample solution: SEC fraction No. 4 ($M_w = 54{,}800$) of a mixture containing 19.6% of copolymer "A", 19.8% "C", 20.5% "E", 19.6% "G", and 20.5% "I" (see Table 9.5), $m_0 = 24.4\,\mu$g, solvent THF. Elution by a gradient methanol/tetrahydrofuran (0-100% in 10 min), flow rate 0.5 ml/min, UV signal 259 nm, 25mAU full scale. The narrow peak with the notation "excl" indicates inertly eluting polymer. The shaded area adjacent to solvent peak "S" varied linearly with sample concentration and was due to improperly retained portions. By courtesy of Friedr. Vieweg & Sohn Verlagsgesellschaft [2]

S/EMA copolymers. The excluded portion of the sample is to be seen in front of the solvent peak.

Similar phenomena occurred when non-fitting combinations of gradients and columns were used, see Fig. 6.4, or when the starting eluent was either too good a solvent or in its polarity not sufficiently different from the column packing. Figure 8.3 demonstrates that a S/AN sample (16.1% AN) was improperly retained when the starting eluent contained more than 50% THF in *iso*-octane.

Figure 6.4 showed that, with S/EMA copolymers, the excluded peak increased with increasing styrene content. This was not only the consequence of the increasing absorptivity of the samples but even more of the decrease in polarity due to the reduced number of EMA units. The latter statement is supported by the tailing of the solvent peak which simultaneously increased with decreasing EMA content. The tailing is caused by polymer eluting together with the sample solvent. With sample "A" (4.7% EMA), this portion increased by more than factor 2 on doubling the sample size, see the two uppermost tracings in Fig. 6.4.

The tailing is related to the hatched area in Fig. 8.2, which indicates a peak immediately following the solvent peak, case (ii). This peak increased linearly with S/EMA content in the injected volume [2]. From this proportionality it was concluded that the peak is caused by improperly retained, sequestered polymer.

Case (iii), i.e. polymer swept through the column by the solvent plug, is the worst one. This portion may easily elute unnoticed. At any rate, quantification is difficult if only optical detectors are available. This problem occurred on injection of S/EMA samples (dissolved in THF) into a starting eluent **A** = *i*Oct + THF (80:20). With copolymers of different EMA content, the area of the supposed solvent peak was (in mAUs per μl sample solution) 200 with sample "A" (4.7% EMA), 161 with "B" (21.2% EMA), and 111 with "C" (32.2% EMA). With proper

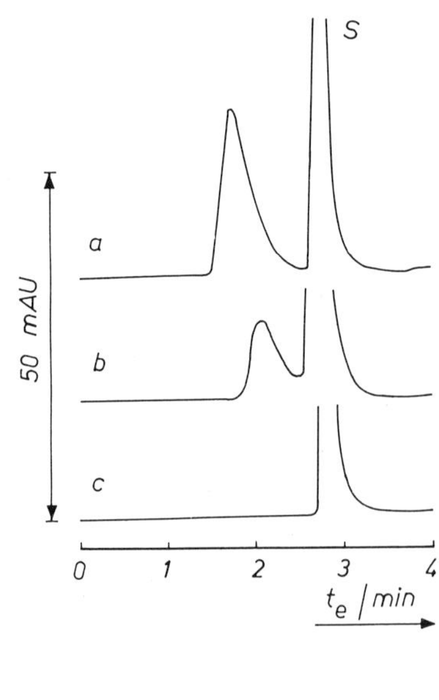

Fig. 8.3. Influence of starting eluent quality on retention of styrene/acrylonitrile copolymers. Sample: mixture of three S/AN copolymers, $m_0 = 43\ \mu g$ in 20 μl THF; injection into *iso*-octane with (a) = 60%, (b) = 55%, or (c) = 50% THF. The peak eluting in the SEC range is from a copolymer with 16.1 wt% AN and $M_n = 325{,}000$ (content in the mixture: 9.6 μg); the other components with more AN were properly retained. Column: 300 × 3.9 mm μ-Bondagel E-1000, $d_p = 10 \mu m$, 50 °C, flow rate 1 ml/min; UV detection at 259 nm, 50 mAU full scale

retention of the polymer, the specific solvent peak was about 60 mAU s/μl (for details, see Ref. [2]).

It must be emphasized that quantification of this kind is scarcely more than a rough approximation because the solvent peak itself is influenced by the age and the storage conditions of the sample solution as well as by the brand of solvent.

The goal must be proper retention of the whole sample. Remedies against any improper retention are (i) active columns, (ii) starting eluents with poor elution strength, and (iii) a high ratio of column to sample volume. Sample volumes in the magnitude of 100 μl should not be injected into analytical columns shorter than 150 mm [2]. This limitation is of importance especially in CCF with pre-fractionation by SEC. 20 μl is the upper limit of injection volume on short columns of, e.g. 60 mm length and moderate polarity.

Evaporative light-scattering detectors and transport detectors are polymer-specific instruments (see Sect. 7.2) which enable any improperly eluting portion of a sample to be noticed easily. With the help of an ELS detector it was confirmed that large injection volumes facilitate sweeping-through of polymer portions [3], see Fig. 9.30.

8.4 Signal Enhancement by Turbidity

Without contribution of adsorption, the elution characteristics of given polymer/eluent systems are, within the limits of experimental error, identical with the solubility characteristics of the systems. This can be seen, e.g. in Figs. 5.6 and 6.15.

In a system of that kind, elution occurs on the verge of precipitation, i.e. in an eluent whose composition cannot guarantee a stable solution. The eluting polymer may be segregated in fine droplets of a gel phase which, in an optical detector, would cause a signal higher than expected from the absorbance of the solute.

This behaviour has been found experimentally in HPLC elution of PS by MeOH/THF or MeOH/DCM gradients [4]. Among the observed effects was the increase of signal size with decreasing flow rate. Figure 8.4 shows the merged plot of the elution at $F = 0.5$ ml/min (left trace), 0.1 ml/min (middle), and 0.03 ml/min (right trace). All other conditions including the sample (a mixture of three PS standards) and the attenuation were identical. Thus, the unbelievable increase in signal size can be seen directly.

In Figure 8.5, the peak area per μg PS, ${}^1A_{259}$, (corrected for $F = 0.5$ ml/min) is plotted vs the square root of inverse flow-rate. The inverse flow-rate is a measure of the period when the polymer samples remain in the solvent/nonsolvent mixture at the verge of precipitation. The diagram indicates a linear increase of signal size with the square root of sample dwell-time. Extrapolation towards zero dwell-time yields $\log {}^1A_{259} = 2.1512$ or ${}^1A_{259} = 141.6$ mAUs/μg. This data corresponds to the specific signal size of PS in DCM (without addition of MeOH) which was 138.4 mAUs/μg at 260 nm (corrected for $F = 0.5$ ml/min). Thus, the higher values in mixed eluents (solvent + nonsolvent) are caused by turbidity.

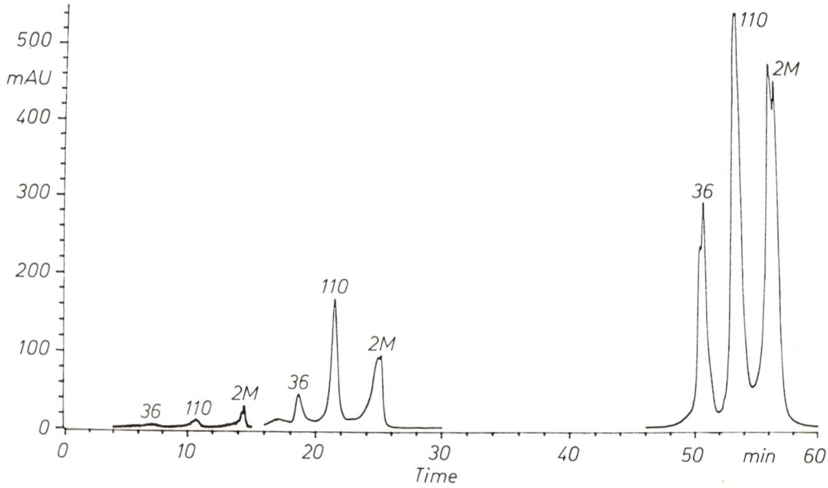

Fig. 8.4. Signal enhancement due to turbidity: merged plot of chromatograms obtained at flow rate 0.5, 0.1, or 0.03 ml/min (from left to right). Sample: mixture of three polystyrene SEC standards (MW 36,000, 110,000, or 2,000,000), 3.2 μg each, $V_O = 24 \mu l$, sample solvent THF. Column RP C18 (60 × 4 mm; $d_O \leqq 5$ nm, $d_P = 5$ μm), gradient: methanol/tetrahydrofuran (36–70% within 17 min). Starting at $t = 1$ min (left), 5 min (middle), or 17 min (right) after sample injection into 100% MeOH and subsequent increase of THF content to 36%; 50′ °C, UV signal at 259 nm, 550 mAU full scale

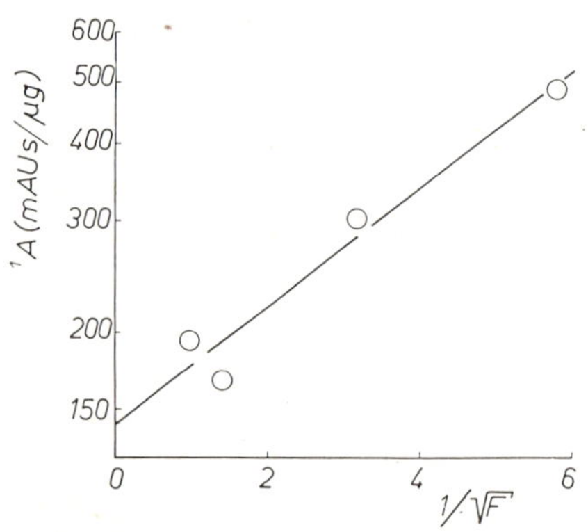

Fig. 8.5. Flow-rate dependence of specific peak area (monitored at 259 nm and normalized to $F = 0.5$ ml/min). Data from Fig. 8.4 and a similar experiment performed at 1.0 ml/min flow rate. By courtesy of Friedr. Vieweg & Sohn Verlagsgesellschaft [4]

The effect increases with dwell-time and depends on MW. At a constant flow rate, a maximum was observed at $M = 470{,}000$. The direct proof of eluate turbidity is the non-zero signal acquired at 400 or 550 nm wavelength where PS itself has no absorption, see Fig. 8.6.

Similar difficulties with sample portions suspended in the mobile phase instead of being precipitated onto the column packing have arosen also with other solubility-based separations, e.g. with temperature-rising elution fractionation (see Sect. 9.14).

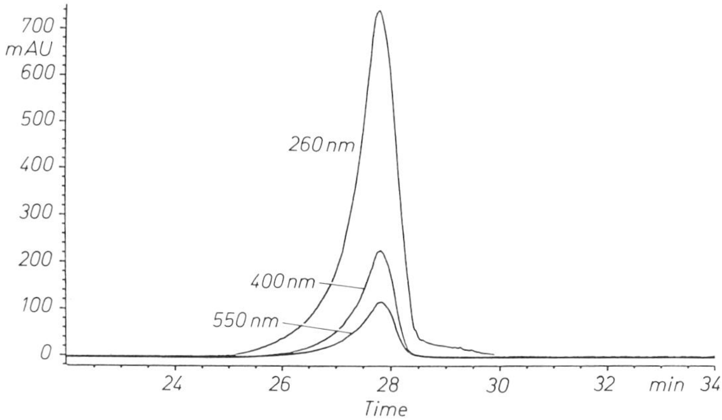

Fig. 8.6. Signal of 300 μg polystyrene (MW 110,000) monitored at the wavelengths indicated. Column: RP C18 (250 × 4.1 mm, $d_O = 9$ nm; $d_P = 10$ μm); 35 °C; gradient: methanol/dichloromethane [20% (0 min), 40% (3 min), 80% (35 min)], flow rate 1 ml/min. By courtesy of Friedr. Vieweg & Sohn Verlagsgesellschaft [4]

8.5 Calibration of Polymer Gradient HPLC

The quantitative evaluation of copolymer chromatograms requires the knowledge of (i) the influence of polymer composition on elution time t_e, (ii) the influence of MW on t_e, and (iii) the influence of composition on detector signal. This knowledge can be gained by calibration with a series of samples graded in composition.

Ideally, each calibrating sample should be homogeneous in composition. In order to obtain samples of narrow CCD, the copolymerization should not be carried out to more than 5% conversion, see Sect. 2.4. But even then a certain heterogeneity in composition must be taken into account. Gradient HPLC of S/MMA copolymers monitored simultaneously by UV at 254 nm (styrene content) and IR at 1730 nm^{-1} (MMA content) revealed about 5% variation in composition [5]. This heterogeneity was found to be essentially of the same order in samples of very low conversion (0.5, 0.7, or 0.8%) as in a sample of 5.4% conversion. It was an example of instantaneous heterogeneity, see Sect. 2.2.

A heterogeneity of about 3% in composition (and 37% in M_n) of a S/MMA standard containing 49.1 mol% styrene was observed by the unintended fractionation of the copolymer in TCM/EtOH on silica at 10 °C [6]. A portion (63%) with low MW and low MMA content was eluted whereas the rest remained in the column until a stronger eluent was used.

The number of samples required for calibration depends on the problem to be investigated. With suitable copolymers, the full range of composition (0 – 100% monomer B) should be considered in basic research. As a rule of thumb, the steps in composition should be 10% or less. More finely graded samples covering a certain range of composition are needed for detailed investigation of important specimens, e.g. non-azeotropic copolymers differing in degree of conversion. Here, the MW of the calibrating samples also should match the MW of the objects of interest.

The elution time of a broad peak is better represented by the first moment of peak area than by its apex. The first moment can be estimated on base of Eq. (8.2) or with the help of suitable software and an electronic integrator. Successful calibration requires the absence of disturbances as described in Sects 8.3 and 8.4.

8.5.1 Influence of Copolymer Composition and Molecular Weight on Elution Time

Since composition and MW influences are interrelated, both effects are at best evaluated simultaneously. This can be done by SEC fractionation of the calibrating samples and gradient HPLC of the fractions. Through universal calibration (see Sect. 3.1) the MW of each fraction can be calculated from sample composition and SEC elution volume. The plot of HPLC elution time (first moment, vide supra) vs $M^{-0.5}$ yielded almost straight lines in every system investigated so far. With suitable systems (polymer, eluent, and chromatographic conditions) the experimental work can be performed on calibrating mixtures of 4–5 individual standards. Provided that the standards are baseline-separated in gradient elution, calibration can be carried out this way much faster without a serious loss of accuracy—but a warning must be given against instability of mixed sample solutions, see Sect. 11.5.

Figure 8.7 is a plot according to Eq. (5.3) of *iso*-octane concentration in *i*Oct/THF gradients vs inverse square root of MW for S/EMA copolymers eluted on either Si50 or CN columns. Among the data displayed are those derived from the chromatograms in Figs. 6.7 and 6.8. As expected, elution from a silica column occurs, in comparison with CN bonded phase, at lower concentration of nonsolvent (i.e. at higher concentration of THF) due to the higher activity of

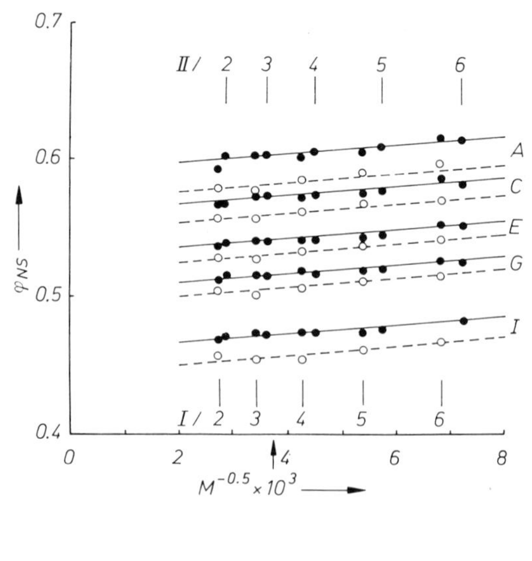

Fig. 8.7. Molecular-weight effect in normal-phase gradient HPLC of *stat*-copoly(styrene/ ethyl methacrylate) samples. Data from Fig. 6.7 (silica column: *circles* and *dashed lines*), from Fig. 6.8 (CN bonded-phase column: *filled circles* and *solid lines*) and from the investigation of another series of SEC fractions (II/2–6) on the same CN bonded phase column. By courtesy of Springer-Verlag [19]

Table 8.1 Molecular-weight effect on retention and solubility: experimental values of parameters C_1 and C_2 in Eq. (5.3)

Sample	Solvent (or eluent component **B**)	Nonsolvent (eluent component **A**)	Measurement (or column)	C_1	C_2	Ref.
PS	DCM	MeOH	RP gradients, glass beads (nonporous)	24.7	3060	[12]
			LiChrospher Si 4000	25.0	3110	[12]
			LiChrospher Si 500	25.1	3070	[12]
			LiChrosorb Si 60	25.2	3070	[12]
			Silica H9010	25.5	3030	[12]
PS	THF	Hex	TT(40 mg/l)	56.4	4370	[13]
		*i*Oct		51.2	4100	[13]
		MeOH		36.8	3100	[13]
		MeOH	TT(10 mg/l)	40.7	3010	[13]
PS	DCM	MeOH	TT	16.8	2430	[15]
S/AN,	DCM	MeOH	TT		4000	[16]
S/AN 16–30% AN	THF + 10% MeOH	Hex, *i*Oct	NP gradient, RP C18		1300	[7]
S/AN, 26.4% AN	THF	*i*Oct	NP gradient, CN bonded ph.		2255	[17]
16.1% AN	THF	*i*Oct	NP gradient,	36.5	2050	[18]
23.0% AN			CN bonded ph.	27.8	2200	[18]
29.2% AN				20.0	2300	[18]
36.4% AN				13.5	2700	[18]
42.5% AN				2.4	2550	[18]
S/EMA						
4.7% EMA	THF	*i*Oct	NP gradient,	57.3	490	[19]
32.2%			Silica	55.6	370	[19]
54.6%				52.7	350	[19]
68.0%				50.3	320	[19]
92.5%				44.8	420	[19]
4.7%	THF	*i*Oct	NP gradient,	59.0	505	[19]
32.2%			CN bond. ph.	56.5	410	[19]
54.6%				53.6	360	[19]
68.0%				51.3	330	[19]
92.5%				47.4	(190)	[19]
4.7%	THF	MeOH	RP gradient,	45.1	970	[20]
32.2%			C18 bond. ph.	56.4	580	[20]
54.6%				64.5	430	[20]
68.0%				69.6	410	[20]
92.5%				78.4	480	[20]
S/EMA	THF	DCE	NP gradient, Silica		510	[21]
S/MEMA	MeOH	*i*Oct	NP gradient,			
25.9%			with 30% THF,	58.4	210	[22]
53.2			CN bonded ph.	52.3	170	[22]
71.2				48.3	200	[22]
87.4				39.2	260	[22]
P(DMA-*b*-MMA), P(MMA-*b*-DMA)			NP gradient,			
	THF	*i*Oct	CN bonded ph.	22.5	4000	[24]
PMMA	THF	*i*Oct	CN bonded ph.	29.4	900	[24]

silica. The slope factors are almost independent of composition. Similar behaviour has been found with S/AN and S/MMA copolymers. Table 8.1 provides a survey on experimental slope factors of the MW effect.

In contrast to this, the RP elution of S/EMA copolymers from a C18 column by a gradient MeOH/THF revealed an influence of polymer composition on the MW effect. With decreasing EMA content, the slope factor approached the value of PS homopolymer.

Except for the high values of PS, S/AN, and block copolymers of DMA with MMA, the slope factors are in the range of 200–1000. A value of 350 can be derived from Fig. 8.7 where the arrow indicates a molecular weight $M = 70{,}000$, which is a reasonable estimate for real copolymers. Around this position a 5%-change in MW corresponds to a change in eluent composition of 0.033% which is less than 1/25 of the change in eluent composition due to 5% difference in EMA content. This illustrates that the sensitivity of gradient HPLC to copolymer composition is much higher than its sensitivity to MW. A similar result was obtained by considering S/AN copolymers in Hex/THF or *i*Oct/THF systems [7].

Necessary corrections for MW dependence provided, the evaluation of tracings from gradient HPLC approaches sufficiently copolymer composition and CC distribution. This has been discussed already with reference to Fig. 3.10. Further support for this statement was found with S/MEMA copolymers by the close agreement between measured composition and the composition calculated by totalling HPLC results (see Sect. 10.4.3.3) and with S/AN copolymers by the correspondence of HPLC results with the results of independent analyses, see Sect. 10.4.1.3.

8.5.2 A General Remark

All published investigations similar to those presented in Figs. 6.7, 6.8, and 6.14 have in common an increase in HPLC retention with decreasing SEC elution volume. Any copolymer specimen whose elution in SEC extended over several SEC fractions showed, with increasing fraction number (i.e. with decreasing MW), a shorter retention in gradient HPLC [8].

One explanation of this effect could be a composition heterogeneity linked to MW distribution. This explanation would raise the question why, in all systems investigated, the content of the more strongly retained monomer should decrease with decreasing MW. This objection not only refers to the variety of monomers which have been used as counterparts to styrene but also to the fact that the phenomenon was independent of whether the monomer ratio in a copolymer was above, below, or equal to that in an *azeotropic copolymer* of the respective system.

As already mentioned, the change in retention is linearly related to $1/\sqrt{M}$, see Fig. 8.7 and Eq. (5.3). It would be strange if this were due to MW-dependent composition changes instead of a direct influence of MW on retention.

An unambiguous solution of this problem can, at the present stage of knowledge, be derived from the behaviour of S/EMA copolymers. Low-MW SEC fractions of these samples are less retained in NP chromatography than high-

MW ones (see Figs. 6.7 and 6.8)..The same sequence in MW holds for retention in RP chromatography (see Fig. 6.14).

Retention of S/EMA copolymers in NP chromatography is mainly due to adsorption. Thus, if one believed in the statement that copolymers in adsorption HPLC were retained independent of MW, one would have to understand the behaviour in Figs. 6.7 and 6.8 as a consequence of reduced EMA content in low-MW fractions. Reversed-phase chromatography of S/EMA copolymers results in an inverted elution order, i.e. samples poor in EMA are longer retained than others rich in EMA. Hence, in RP chromatography the low-MW fractions should be eluted later than the high-MW ones-but, in fact, they are eluted *earlier*. Thus, with S/EMA copolymers the MW effect in gradient HPLC is beyond any doubt and proved to be substantial even for an adsorption mechanism.

Dual-detection gradient HPLC (similar to the investigations shown in Fig. 9.5) of SEC fractions would reveal whether or not the composition of copolymer standards changes with MW; thus it should be a promising method for clarifying the problem of MW influence in gradient HPLC.

8.5.3 Influence of Copolymer Composition on Detector Response

In well-behaved chromatograms, the peak area A is the true measure of the amount injected. Changing sample size must cause the peak area to change linearly with input. This basic condition is fulfilled in SEC of polymers, of course. The present discussion is directed to the behaviour of polymers in gradient HPLC.

8.5.3.1 S/AN Copolymers. Figure 8.8 shows a plot of UV peak area vs sample mass for a copolymer containing 30% AN. The response is linear, the straight line has a slope $f = 15.3$ (in $mm^2/\mu g$) which, under the conditions employed, is the calibration factor for samples of that composition. Specimens with less AN content have steeper calibration lines because at 259 nm UV absorption is caused by styrene units.

Figure 8.9 reveals that the slope f does not vary linearly with styrene content. This is partly due to a sequence-length effect which leads to hypochromism in

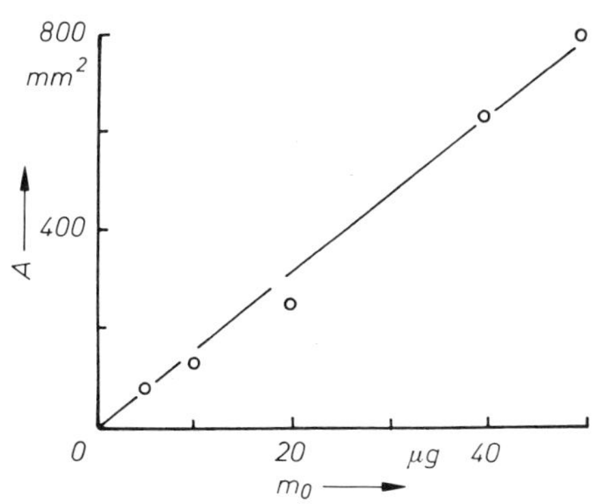

Fig. 8.8. Peak area (UV signal at 259 nm) vs sample mass for a *stat*-copoly(styrene/acrylonitrile). Sample: 30 mass% AN, $M_n = 71{,}000$, $m_o = 5 - 50\,\mu g$ in $V_o = 50\,\mu l$; sample solvent: THF. Column RP C18 (150 × 4.6 mm; $d_o \leqq 10$ nm, $d_P = 10\,\mu m$); gradient *iso*-octane/tetrahydrofuran, multilinear from 60 to 90%, flow rate 1 ml/min, 50 °C. By courtesy of IUPAC [7]

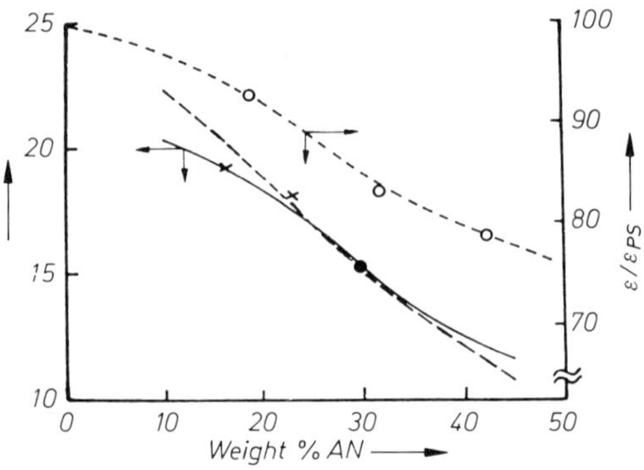

Fig. 8.9. Hypochromic effect with *stat*-copoly(styrene/acrylonitrile) samples. o: Normalized extinction at 261 nm vs AN content (S/AN in THF, data from Ref. [9]; response factor dA/dm_0 read from Fig. 8.8; x: experimental response factors at other copolymer composition. Solid line: experimental calibration curve, dashed line: curve from Ref. [9], normalized to fit point "•". By courtesy of IUPAC [7]

styrene absorption via adjacent AN units [9, 10]. The figure shows the result from Fig. 8.8 ($f = 15.3$ at 30% AN, full dot) and (indicated by crosses) the calibration factors of two other samples of lower AN content. This data has been also obtained by runs using the system *i*Oct/THF.

The upper dashed line and the circles (from the work of Brüssau and Stein [9]) indicate a hypochromic effect. The lower dashed line is drawn on basis of the measured value $f = 15.3$ (for 30% AN, vide supra) and the hypochromic effect represented by the upper dashed curve. The curve shows the dependence of calibration factor f on copolymer composition in sufficient agreement with chromatographic results.

8.5.3.2 S/EMA Copolymers. If peak area A varies linearly with m_0, the dependence of UV signal on copolymer composition (similar to Fig. 8.9) can also be derived from the specific peak area (in mAUs/μg) of suitable calibrating samples. Figure 8.10 shows the peak area per μg vs styrene content of S/EMA copolymers. The circles are from Fig. 6.5 where 5 μl containing 5 μg each were injected into a starting eluent *i*Oct+THF (80:20). The full dots show data calculated from a similar set of chromatograms measured on injection of 10 μl sample solution containing 10 μg copolymer each. Here, the starting eluent was 100% *i*Oct. Both sets of data show the same increase of peak area with composition. The straight line yields the relation [2]

$$^1A_{259} = 1.29 \cdot x \quad (8.7)$$

where $^1A_{259}$ is the peak area at 259 nm wavelength (in mAUs) and x (in mol%) is the styrene content of copolymers. From this equation, the specific peak area of PS homopolymer is calculated to be 129 mAUs per μg which is in reasonable agreement with the corresponding data reported in Sect. 8.4. Hence, the signals of S/EMA copolymers with high or intermediate EMA content, which determine the slope factor in Eq. (8.7), are definitely not enhanced by turbidity of the eluate.

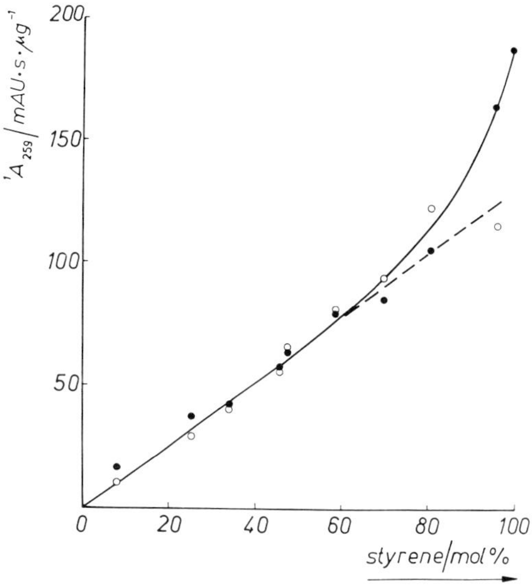

Fig. 8.10. Dependence of specific peak area (UV signal at 259 nm per μg sample amount) on styrene content of *stat*-copoly(styrene/ethyl methacrylate) samples. Column: CN bonded-phase, gradient *iso*-octane/tetrahydrofuran. *Circles* from Fig. 6.5 (injection of 5 μl into *i*Oct/THF 80:20), *filled circles* from Fig. 8.1 (10 μl into 100% iOct). By courtesy of Friedr. Vieweg & Sohn Verlags gesellschaft [2]

In Fig. 8.10, the uppermost right dot (PS homopolymer) and the adjacent value of a S/EMA sample containing 95.7% S are obviously too high. Thus, signal enhancement by turbidity is possibly peculiar to polystyrene and closely related materials.

8.5.3.3 Further Copolymers. With *stat*-copoly(*S*/butadiene) samples, Sato et al. [11] found a better reproducibility of peak area (254 nm UV absorbance) on columns packed with cross-linked PAN than on silica columns. (With the latter, a gradient Hex/THF was used instead of Hex/TCM with PAN packings.) The influence of the column packing material can be understood as the consequence of improper retention or elution, see Sect. 8.3.

In principle, the evaluation of peak area as a function of sample composition, see Fig. 8.10, can also be carried out using mixed standards. In an investigation with S/MMA copolymers, mutual disturbance of the last eluting components of a mixed sample was observed in Hex/THF gradients on silica columns but, on cross-linked PAN packings, correct increase of peak area with styrene content in Hex/TCM gradients was found [5].

At 254 or 259 nm, absorptivity is due to styrene content in copolymers with AN, MMA, EMA, or MEMA as comonomers. At 230 nm, MEMA units contribute substantially to absorption. This has been already mentioned in Sect. 5.6 and enabled monitoring of copolymers rich in MEMA (see Fig. 9.22) as well as, in a limited range of composition, evaluation of S/MEMA recordings assuming detector response to be independent of composition (see Sect. 10.4.3.1).

8.6 References

1. Wiese A, Dehmer B, Dörr T, Höschele G (1984/4) Hewlett-Packard J 35: 26
2. Glöckner G (1987) Chromatographia 23: 517
3. Augenstein M, Stickler M (1990) Makromol. Chem. 191: 415
4. Glöckner G, Schmutzler S, Engelhardt H, Schultz R (1988) Chromatographia 25: 983
5. Sato H, Takeuchi H, Tanaka Y (1986) Macromolecules 19: 2613
6. Mori S, Uno Y (1987) Analyt Chem 59: 90
7. Glöckner G (1983) Pure Appl Chem 55: 1553
8. Sato H, Mitsutani K, Shimizu I, Tanaka Y (1988) J Chromatogr 447: 387
9. Brüssau RJ, Stein DJ (1970) Angew Makromol Chem 12: 59
10. Garcia-Rubio LH, Hamielec AE, MacGregor JF (1982) ACS Symposium Series 197: 151
11. Sato H, Takeuchi H, Tanaka Y (1985) Int Rubber Conf, Kyoto [18B15]: 596
12. Schultz R, Engelhardt H (1990) Chromatographia 29: 205
13. Glöckner G, Meißner C unpublished, measured for Ref [14]
14. Glöckner G (1988) Chromatographia 25: 854
15. Schultz R (1989) Thesis, University of Saabrücken
16. Glöckner G, Francuskiewicz F, Müller S (1975) Faserforsch Textiltechnik 26: 287
17. Glöckner G, van den Berg JHM, Meijerink NLJ, Scholte ThG (1986) In: Kleintjens LA, Lemstra PJ (eds) Integration of fundamental polymer science and technology, Elsevier, Barking, UK p 95
18. Glöckner G computations performed in preparing the manuscript of Ref [17]
19. Glöckner G, Stickler M, Wunderlich W (1987) Fresenius Z Anal Chem 328: 76 (computations performed on data of mixture M1)
20. Glöckner G, Stickler M, Wunderlich W (1988) Fresenius Z Anal Chem 330: 46 (computations performed on data of mixture M1)
21. Danielewicz M, Kubin M (1981) J Appl Polym Sci 26: 951
22. Glöckner G computations performed in preparing the manuscript of Ref [23]
23. Glöckner G, Stickler M, Wunderlich W (1989) J Appl Polym Sci 37: 3147
24. Augenstein M, Müller MA (1990) Makromol Chem 191:2151

9 Separation of Copolymers by Composition through Gradient High-Performance Liquid Chromatography

9.1 Statistical Copolymers from Styrene and Acrylonitrile

Azeotropic copolymers of this system (*stat*-copoly(S/AN); $r_s = 0.41$; $r_{AN} = 0.04$) contain 24 mass% AN or 61.9 mol% S. Products of about this composition are commercially available. Furthermore, S/AN is of importance for high-impact ABS composite polymers. S/ANs with less than 40 mass% AN are soluble in common organic solvents whereas copolymers richer in AN require DMF or DMSO which are solvents even for polyacrylonitrile.

S/AN copolymers have been fractionated by composition through gradient elution with CHx/MEK [1,2] or (Tol + PrOH 50:50)/ DMF [3], through step gradients of ethylene cyanohydrin/ethylene carbonate [4], or through *Baker-Williams fractionation* with Hex/DCM gradients [5].

HPLC separation of S/AN copolymers by composition was first performed on columns packed with either silica or C8 bonded-phase material of $d_P = 10\,\mu m$ grain size and $d_0 = 10\,nm$ pore diameter [6]. The injection volume was $50\,\mu l$ containing $20\,\mu g$ sample polymer in THF solution, the gradient was Hex/THF. It was found that (i) a given copolymer is eluted in a constant eluent composition irrespective of whether it was injected alone or in a mixture with other samples, (ii) the eluent composition at peak elution corresponds to the precipitation threshold of the respective sample, and (iii) the peak area doubles on doubling the amount injected. The investigation was continued with *i*Oct/THF gradients and a variety of additional stationary phases [7–13]. Among the results were those displayed in Figs. 5.8 and 10.4. The data of the copolymers investigated are compiled in Table 9.1. All HPLC attempts made so far to separate S/AN copolymers by composition yielded retention times increasing with AN content of the samples.

In another approach to separate these samples, a CN bonded-phase column (50×4.0 mm; $d_0 = 6$ nm; $d_P = 10\,\mu m$) was used together with a gradient starting with a 70:30 (v/v) mixture of heptane and DCM. The solvent **B** was AcN + DCE (40:60) whose concentration was linearly increased from 3 to 100% within 13 min [14].

Chromatographic cross-fractionation of S/AN copolymers is dealt with in Sects. 10.3.1 (mixtures of model copolymers) and 10.4.1 (analysis of a commercial product).

Table 9.1 S/AN copolymers (M_s: 104.144, M_{AN}: 53.062)

Code	A	B	C	D	E
$100w_{AN}$	16.1	23.0	29.2	36.4	42.5
$10^{-3} M^a$	325	480	510	380	340
Convers. (%)	3.6	6.7		7.3	10.3

[a] by osmosis

9.2 Statistical Copolymers from Styrene and Methyl Acrylate

Azeotropic copolymers of this system (*stat*-copoly(S/MA); $r_s = 0.75$, $r_{MA} = 0.18$) contain 20.1 mass% MA or 76.6 mol% S units. Mixtures of the samples A, B, and C (listed in Table 9.2) have been separated by composition on silica columns ($d_P = 10$ or $15\,\mu m$) through Tetra/MeAc gradients with MeAc content increasing by 0.8%/min [15]. Figure 9.1 shows a chromatogram achieved by this technique from the mixture of the three specimens.

Retention increased with increasing MA content (equal to increasing polarity) of the sample polymer. This behaviour fits the normal-phase chromatographic system used. In spite of the decaying activity of the columns, the composition distribution of sample B could be evaluated. (The authors observed contamination of the silica packing by decomposition products of tetrachloromethane and, thus, decreasing column activity. Similar difficulties have been reported recently also with Hex/TCM gradients on silica columns [16].)

S/MA copolymers have been also investigated on a silica column by gradient Tol/MEK with MEK content increasing by 2.4%/min. Separate injections (10 μg in 10 μl Tol/MEK 98:2) of samples D-G (see Table 9.2) yielded retention times increasing with MA content [17]. No information has been given about temperature control. Figure 9.2 shows that samples D and E, whose composition

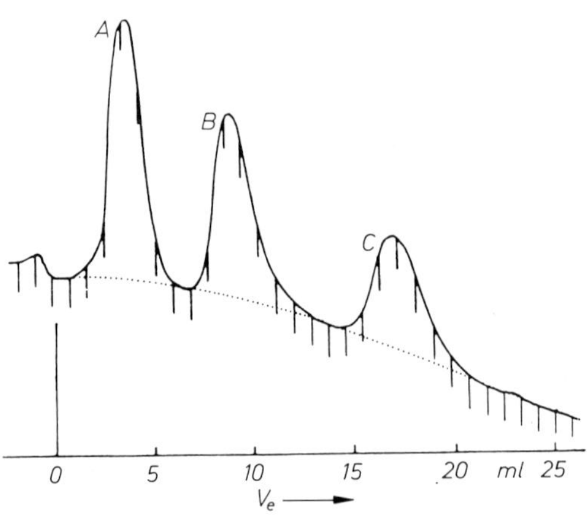

Fig. 9.1. Separation of the mixture of three *stat*-copoly(styrene/methyl acrylate) specimens on a silica column (600 × 7.5 mm; $d_o = 5$ nm; $d_p = 15\,\mu m$) through a gradient tetrachloromethane/methyl acetate (7–35% in 35 min), flow rate 1 ml/min. Copolymers A, B, and C, data see Table 9.2. From Ref. [15] with permission

Table 9.2 S/MA copolymers (M_s: 104.144, M_{MA}: 86.088)

Code	A	B	C	D	E	F	G
$100 w_{MA}$	41.9	52.6	74.4	15.8	25.0	37.1	54.1
$10^{-3} M$	261[a]	276[a]	302[a]	78.3[b]	93.6[b]	119.3[b]	133[b]
Convers. (%)	13.4	9.8	9.2	<10	<10	<10	<10

[a] by osmosis
[b] SEC (number average)

is close to the azeotropic one, eluted in narrow peaks whereas the peaks of specimens F and G became broader with increasing distance from the azeotropic point. The CCD of these samples, derived from the elution profile, fits the calculated ones even better than in an example given in Ref. [15]. This improvement may be due partly to reduction in peak broadening by intra- and extra-column effects. In addition, in Fig. 9.2 an evaporative light-scattering (ELS) detector (see Sect. 7.2.2) was used, which is capable of monitoring the whole copolymer, whereas in Ref. [15] a UV detector was employed which monitors styrene units only. Furthermore, the use of an ELS detector allowed separation with a gradient of toluene and MEK, which both absorb UV radiation in the region where the polymer can be detected.

Recently, separation of S/MA on silica through Hp/**B** gradient (10–100% **B** in 18 min) has been reported [18]. Eluent **B** was DCM + 4% MeOH. Contribution of adsorption to retention was observed on silica, glass, and polar bonded-phase columns (diol > CN > NH_2) but not on C18. Inversion of elution order was observed in reversed-phase chromatography with MeOH/DCM gradients on C8, C18, or CN columns. Here, solubility characteristics and solubility line almost coincided.

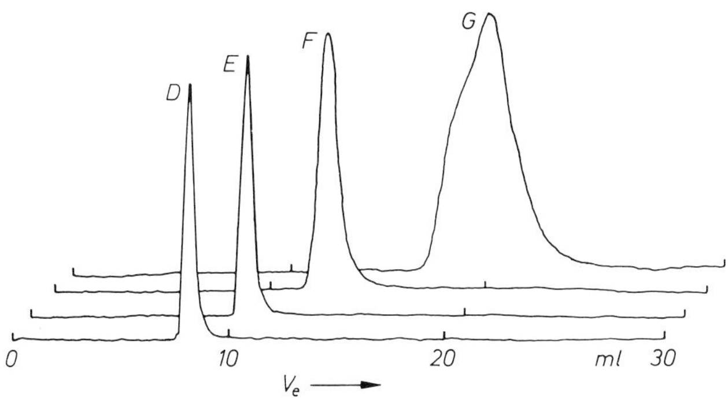

Fig. 9.2. Gradient elution of *stat*-copoly(styrene/methyl acrylate) samples on a silica column (250 × 4.6 mm; $d_o = 6$ nm; $d_p = 5\,\mu$m), monitored by an evaporative light-scattering detector. Gradient: toluene/methyl ethylketone (2 – 50% in 20 min), flow rate 1 ml/min. Copolymers D-G, data see Table 9.2. From Ref. [17] with permission

9.3 Copolymers from Styrene and Methyl Methacrylate

9.3.1 Statistical Copolymers

Azeotropic copolymers of this system (*stat*-copoly(S/MMA); $r_s = 0.53$, $r_{MMA} = 0.49$) contain 47 mass% MMA or 52 mol% S units. The copolymers are soluble in many common organic solvents. Alkane hydrocarbons and lower alcohols act as precipitants. S/MMA samples investigated by gradient HPLC and their characteristics are listed in Table 9.3.

Separation by composition has been achieved in numerous eluents. Most of them represent, together with the respective columns, proper normal-phase systems with gradients increasing in polarity and a polar stationary phase. Silica columns have been used frequently. On a column of this kind (250 × 6 mm, $d_0 = 6$ nm, $d_P = 9$ μm) and with a gradient dichloroethane/THF (**B** exponential from 3 to 20%) the mixture of samples C, D, and E was separated [19]. The shape of the gradient was recorded by means of a differential refractometer, whereas the elution of the copolymers was monitored by UV absorption. Before each analysis, the column was flushed with at least ten column volumes THF. This way, perfect reproducibility was achieved during a period of several months [19].

Figure 9.3 shows the separation of the mixture of samples A-G on a silica column through a gradient *i*Oct/(THF + 10% MeOH). The peaks were eluted at a flow rate of 0.3 ml/min [20]. Similar chromatograms were obtained by CHx/THF [21] or Hex/THF gradients.

Table 9.3. S/MMA copolymers (M_s: 104.144, M_{MMA}: 100.114)

Code	A	B	C	D	E	F	G
$100 w_{MMA}$	11.4	23.8	37.0	49.5	64.0	76.2	88.5
$10^{-3} M^a$	160	250	150	185	235	220	220
Code	H	I	J	K	L	M	N
$100 x_{MMA}$	33.7	42.6	51.3	57.9	58.5	73.5	84.8
$100 w_{MMA}$	32.8	41.6	50.3	56.9	57.5	72.7	84.3
Code	O	P	Q	R			
$100 w_{MMA}$	14.1	24.8	50.3	76.2			
$10^{-3} M^b$	140	188	127	122			
Convers. (%)	10.9	2.9	17.0	32.8			
Code	S	T	U	V	W	H3	
$100 w_{MMA}$	18.4	32.1	43.0	61.1	75.3	67.2	
$10^{-3} M^c$	149	184	138	218	273		
Convers. (%)	0.7	0.8	0.5	5.4	8.9	97.6	
Code	X1	X2	X3	X4	X5		
$100 w_{MMA}$	21.0	31.2	45.3	63.7	75.3		
$10^{-3} M^c$	139	124	119	135	243		
Convers. (%)	5.0	6.9	6.1	7.6	7.6		

[a] by light scattering; [b] by osmosis; [c] SEC (weight average)

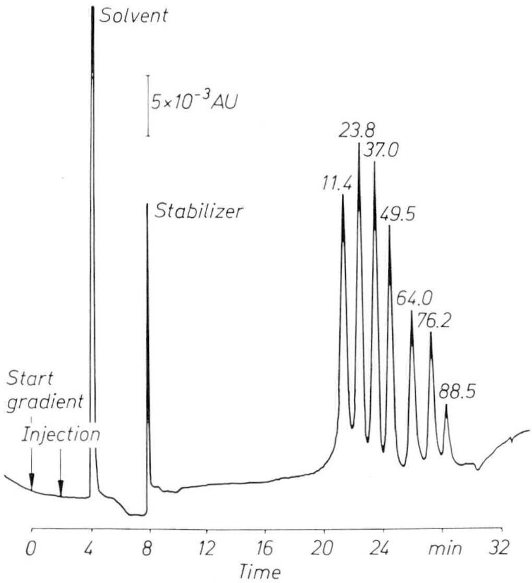

Fig. 9.3. Separation of the mixture of seven *stat*-copoly(styrene/methyl methacrylate) specimens on a silica column (150 × 4.6 mm; $d_o = 6$ nm; $d_p = 5\ \mu$m) by gradient HPLC at 50 °C. Eluent **A**: *iso*-octane, eluent B: tetrahydrofuran + 10% methanol; multilinear gradient from 10% **B** (0 min) to 50% **B** (8 min), 80% **B** (10 min), and 100% **B** (11 min), flow rate (0 – 9.9 min) 1 ml/min and (from 10.0 min onwards) 0.3 ml/min. Sample composition (mass% MMA) indicated, for further data see Table 9.3, samples A-G. UV signal at 259 nm. By courtesy of Elsevier Science Publishers [20]

The mixture of the S/MMA copolymers K, M, and N was separated on a silica column (50 × 4.6 mm, $d_0 = 3$ nm, $d_P = 5\ \mu$m) by a gradient trichloromethane/ethanol [22]. In fact, eluent **A** was TCH + EtOH (99:1) and eluent **B** TCM + EtOH (95.5:4.5). The flow rate was 0.5 ml/min, the gradient program 0–100% **B** in 15 min, and the column temperature 20 °C. 50 μl were injected of a sample solution in eluent **A** containing equal amounts of the three copolymers (in total 25 μg). Similar results could be obtained at 30 and 80 °C with 100 μl injections [23]. In the latter case, four samples could be separated (H, I, J, and K of Table 9.3).

Sato et al. [16] investigated S/MMA copolymers by gradient HPLC on a variety of column packings (including highly cross-linked polymers) and monitored the elution by UV (254 nm, sensitive to styrene) and IR detection (1730 cm^{-1}, sensitive to MMA). Figure 9.4 shows the elution pattern obtained on a column packed with cross-linked PAN and a gradient Hex/TCM. From the peak-height ratio UV/IR, the actual composition of the copolymer could be derived. Figure 9.5 indicates the chemical heterogeneity of calibrating samples, which had been mentioned already in Section 8.5.

On basis of this calibration, the CCD of four fractions from sample V could be evaluated, see Fig. 9.6 and also that of copolymers from the same monomer batch as sample V but polymerized to higher conversion, see Fig. 9.7. The experimental curve of the sample at 98% conversion almost fits to the calculated one and is even narrower than the latter. Taking into account the chromatographic peak broadening, which can be estimated by comparison of the experimental and calculated patterns in Fig. 9.6 or the curves in Figs. 9.7a and 9.7b, the

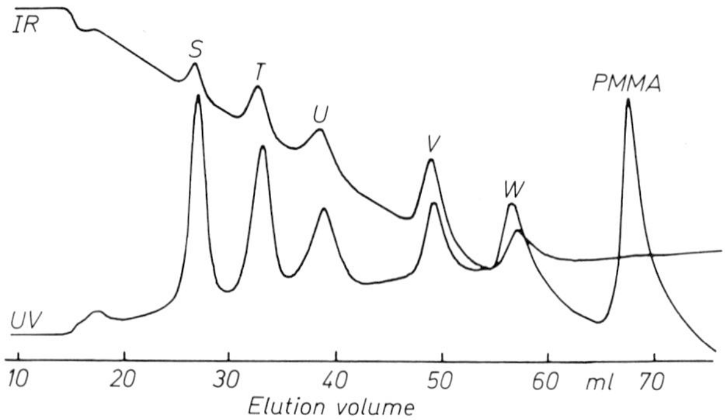

Fig. 9.4. Separation of the mixture of five *stat*-copoly(styrene/methyl methacrylate) specimens and PMMA homopolymer on crosslinked polyacrylonitrile. Column 550 × 7.5 mm; gradient hexane/chloroform, monitored by UV (254 nm) and IR detector (1730 cm^{-1}). Samples S, T, U, V, and W, data: Table 9.3. From Ref. [16] with permission

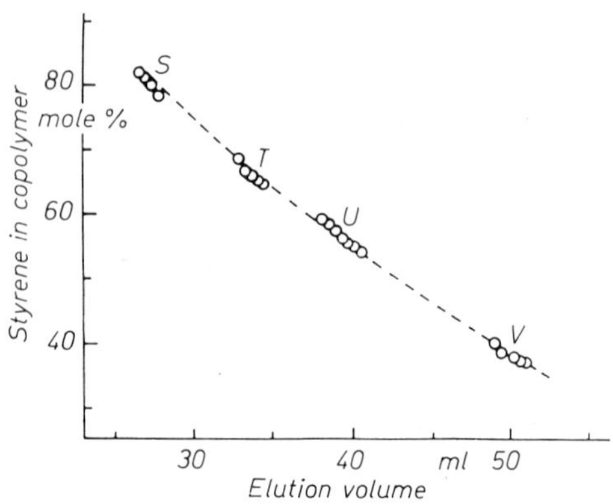

Fig. 9.5. Actual compositon of *stat*-copoly(styrene/methyl methacrylate) samples S, T, U, and V (see Table 9.3), estimated from the height ratio of UV and IR recordings, see Fig. 9.4. From Ref. [16] with permission

experimental CCD of the 98%-conversion sample is, in fact, probably narrower than the calculated one.

Chromatographic cross-fractionation of a high-conversion S/MMA copolymer is reported in Sect. 10.4.2.

S/MMA copolymers can be well separated by composition through NP chromatography. The resolution is sensitive to the activity of the stationary phase; it diminishes to virtually zero if non-fitting combinations of gradient and column are used (see Fig. 6.10). Proper RP systems are also effective in separating S/MMAs. This has been demonstrated by Teramachi [24] in the investigation of a given sample mixture on a CN bonded-phase column by a gradient CHx/THF, see Fig. 9.8, or on a C18 column by a gradient AcN/THF, see Fig. 9.9. Excellent separation was also achieved on a phenyl bonded-phase column (150 × 3.9 mm; $d_0 = 10$ nm, $d_p = 5$ μm) through a gradient AcN/THF (5–55% in 15 min). In all these investigations, 120 μl of a THF solution containing 30 μg of the samples

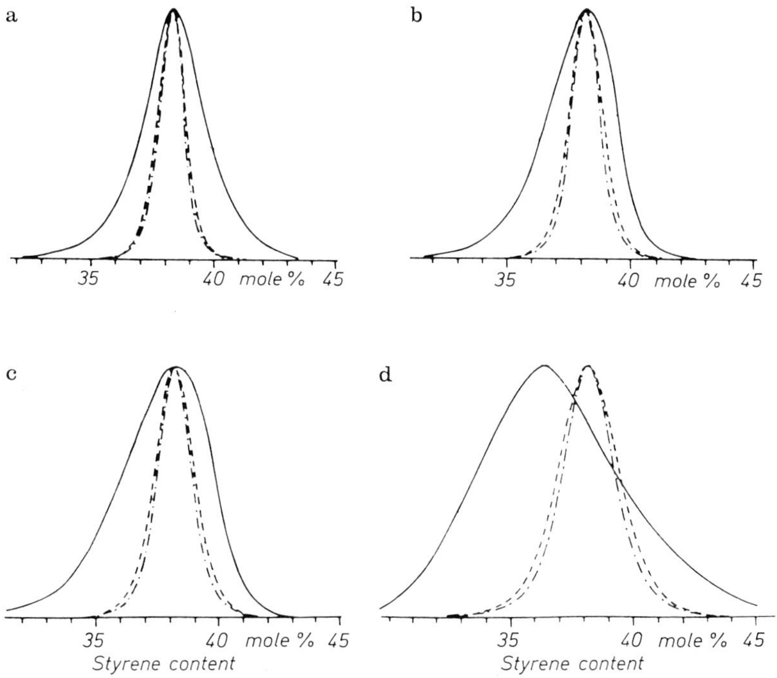

Fig. 9.6 a–d. Chemical composition distribution of four SEC fractions from *stat*-copoly(styrene/methyl methacrylate) estimated from normal-phase gradient HPLC. Chromatographic conditions see Fig. 9.4; *dashed curves*: calculated under the assumption of termination by recombination; curves indicated by *dash/point* combination: ditto, but termination by disproportionation. Molecular weight of the fractions ($10^{-3}M$): (**a**) 350; (**b**) 295; (**c**) 187; (**d**) 99. From Ref. [16] with permission

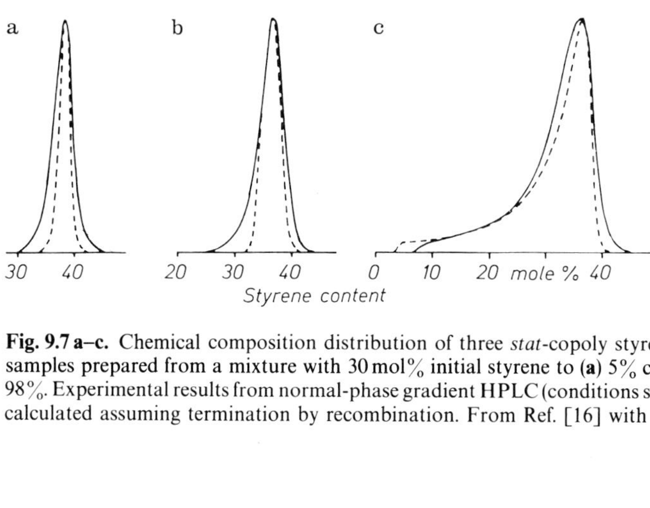

Fig. 9.7 a–c. Chemical composition distribution of three *stat*-copoly styrene/methyl methacrylate) samples prepared from a mixture with 30 mol% initial styrene to (**a**) 5% conversion, (**b**) 42%, or (*c*) 98%. Experimental results from normal-phase gradient HPLC (conditions see Fig. 9.4). *Dashed curves* calculated assuming termination by recombination. From Ref. [16] with permission

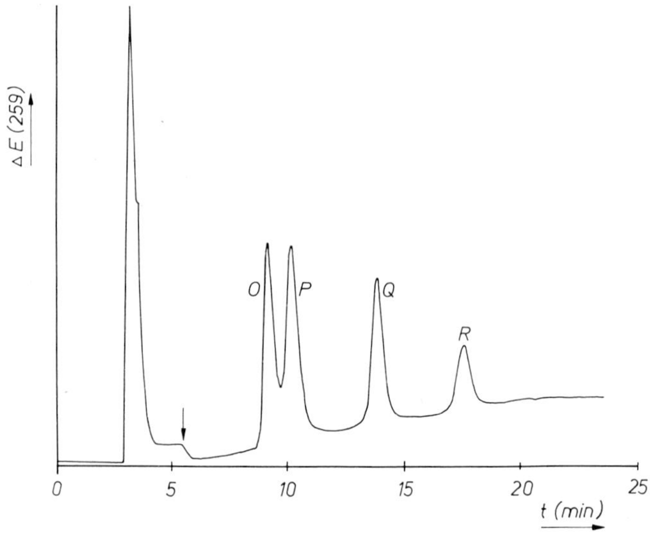

Fig. 9.8. Separation of the mixture of four *stat*-copoly(styrene/methyl methacrylate) specimens by normal-phase gradient HPLC on a CN bonded-phase column (300 × 3.9 mm) at 30 °C. Gradient cyclohexane/tetrahydrofuran (10 – 60% in 15 min), flow rate 1 ml/min; sample data: Table 9.3. From Ref. [24] with permission

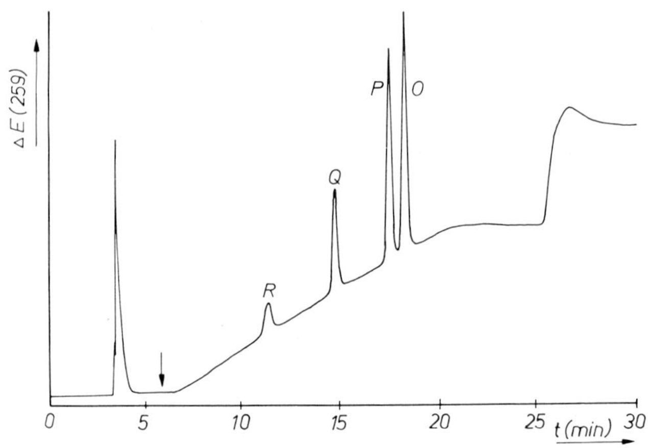

Fig. 9.9. Separation of the same mixture as in Fig. 9.8 by reversed-phase gradient HPLC on a C18 bonded-phase column (250 × 4.6 mm) at 30 °C. Gradient acetonitrile/tetrahydrofuran (5 – 55% in 15 min), flow rate 1 ml/min. From Ref. [24]with permission

O–R (see Table 9.3) had been injected. The column temperature was 30 °C, the flow rate 1 ml/min.

RP separation of S/MMA copolymers has been also achieved through AcN/DCM gradients on columns packed with cross-linked PS [25], samples X1 – X5 (see Table 9.3).

9.3.2 Block Copolymers

Poly(styrene-*b*-methyl methacrylate) samples were investigated by gradient HPLC using TCM/EtOH (0 or 1%–4.5%, 0.233% EtOH/min) on silica column (50 × 4.6 mm, $d_0 = 3$ nm, $d_P = 5\ \mu$m) at 30 °C [26]. Rather large amounts (100 μg in 100 μl of the starting eluent) were injected and the chromatograms monitored by UV detection at 254 nm. The samples were polymerized by free radical reaction through multifunctional peroxides at either 65 °C or 75 °C for MMA or styrene blocks, respectively, and had broad distributions in MW as well as in composition. Two main components could be isolated: one with both low MMA content and MW, the other with both high MMA and MW.

9.4 Statistical Copolymers from Styrene and Ethyl Acrylate

Azeotropic copolymers of this system (*stat*-copoly(S/EA); $r_s = 0.80$, $r_{EA} = 0.20$) contain 19.4 mass% EA or 80 mol% S units. The samples A–D listed in Table 9.4 were polymerized in solution (25 ml monomer mixture in 50 ml benzene) at 60 °C. After a reaction time of 10 h, the solutions were cooled to about 0 °C and most of the benzene evaporated under reduced pressure. Then the copolymer was precipitated by pouring the remaining liquid into 1-hexanol. The polymer yield was between 10 and 20%.

A mixture of the S/EA copolymers A–D was separated on a silica column at 60 °C by gradient TCM/EtOH (1–7% in 30 min) [27]. Retention increased with increasing temperature; at 40 °C, sample A (31.4 mol% EA) eluted unretained in the interstitial volume. The behaviour of S/EA copolymers was similar to that of S/EMA (see Fig. 9.10) under corresponding conditions.

9.5 Statistical Copolymers from Styrene and Ethyl Methacrylate

Azeotropic copolymers of this system (*stat*-copoly(S/EMA); $r_s = 0.49$; $r_{EMA} = 0.40$) contain 48.2 mass% EMA or 54.1 mol% S units.

A sample containing 47.1 mol% (49.4 mass%) EMA was mentioned in a paper dealing with the separation by composition of S/MMA copolymers [19].

Table 9.4. S/EA copolymers (M_s: 104.144, M_{EA}: 100.114)

Code	A	B	C	D
$100x_{EA}$	31.4	47.4	63.3	79.3
$100w_{EA}$	30.6	46.6	62.4	78.6
	10–20% conversion; MW (SEC weight average): 100–300 × 10^3			

(Separation by composition of S/EMA copolymers is not reported in Ref. [19].) The sample mentioned was fractionated according to MW. Five fractions of essentially the same content of styrene were separately injected into a column (250 × 6 mm) packed with silica ($d_0 \approx 5$ nm, $d_P = 9\ \mu$m) and eluted by a gradient DCE/THF (3–20%, exponential). The author's statement that "the eluent composition at the respective peak maxima... was independent of MW down to about 1×10^5" is contradicted by the fact that plotting the experimental results according to Eq. (5.3) yields a slope factor $C_2 = 510$.

Table 9.5 lists the characteristics and sample codes of S/EMA copolymers which have been separated by composition through NP or RP gradient HPLC [28–30], see Sect. 6.3. In S/EMA copolymers, the constituting units are rather well balanced in polarity. This facilitates the application of both NP and RP separation modes with inversion of elution order, see Ref. [30].

On the other hand, the separation according to composition of S/EMA copolymers is more difficult than that of S/MMA samples. This drawback and a pronounced MW effect are the reasons why, with mixtures of samples A–I listed in Table 9.5, baseline separation could be obtained only after prefractionation by SEC.

The elution patterns of *i*Oct/THF (20–80%) gradient separations on columns packed with either silica or CN bonded-phase are shown in Figs. 6.7 and 6.8, respectively, whereas the RP separation by a MeOH/THF gradient on a RPC18 column is displayed in Fig. 6.14. For the NP systems, the effect of sample composition and MW on eluent composition at the first moment of the respective peaks is presented in Fig. 8.7 in the context of calibration recommedations. As can be read from Table 8.1, a slope factor $C_2 = 510$ (vide supra) in DCE/THF fits the data for S/EMA in other systems.

Recently, the separation of S/EMA model mixtures by gradient TCM/EtOH in normal-phase chromatography was reported [27]. The result obtained with specimens J–M (Table 9.5) is illustrated by Fig. 9.10 which also shows the pronounced effect of temperature on retention.

Table 9.5. S/EMA copolymers (M_s: 104.144, M_{EMA}: 114.140)

Code	A	B	G	D	E	F	G	H	I
$100w_{EMA}$	4.7	21.2	32.2	43.4	54.6	56.3	68.0	76.2	92.5
$10^{-3}M^a$	51.6	56.5	63.1	68.9	65.2	65.8	83.6	115	61.6
Code	J	K	L	M					
$100x_{EMA}$	30.9	49.8	69.6	84.5					
$100w_{EMA}$	32.9	52.1	71.5	85.7					
	(10–20% conversion)								

[a]SEC (weight average)

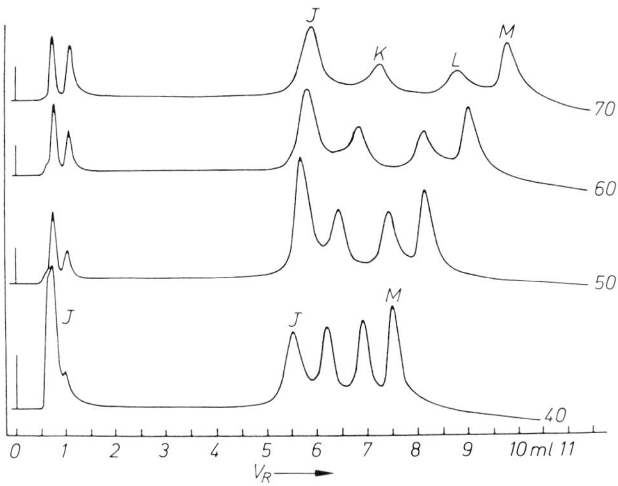

Fig. 9.10. Separation of the mixture of four styrene/ethyl methacrylate copolymers on a silica column by gradient chloroform/ethanol (1–7% in 30 min). Eluent **A** TCM + EtOH (99:1), **B**: TCM + EtOH (93:7); flow rate 0.5 ml/min; column 50 × 4.6 mm, $d_o = 3$ nm, $d_p = 5\,\mu$m; samples: J – M in Table 9.5; $m_o = 25\,\mu$g, $V_o = 25\,\mu$l; temperature indicated (40–70 °C). From Ref. [27] with permission

9.6 Copolymers from Styrene and *t*-Butyl Methacrylate

9.6.1 Statistical Copolymers

Azeotropic copolymers of this system (*stat*-copoly(S/TBMA); $r_s = 0.56$; $r_{TBMA} = 0.60$) contain 60.05 mass% TBMA or 47.6 mol% S units. Table 9.6.1 gives a survey of the samples investigated.

In a proper NP system formed by the combination of a silica column and a gradient *i*Oct/THF, the specimens A–D were, in a first approximation, retained to almost the same degree. This was unexpected since S/MMA and S/EMA copolymers of differing composition could be separated well by NP gradient HPLC. (Similarly, the more difficult separation of S/BA in comparison with S/MA copolymers was recently observed by Dutch researchers [18]) With S/TBMA copolymers, the only difference was a slight decrease in retention with increasing content in methacrylate units. In contrast to this, the NP retention of S/MMA and S/EMA copolymers increased with methacrylate content. Figure 9.11 is the synoptic presentation of the elution characteristics of methacrylate copolymers with different ester groups. The steepness of the characteristics

Table 9.6.1. Statistical copolymers, S/TBMA (M_s: 104.144, M_{TBMA}: 142.192)

Code	A	B	C	D
$100 w_{TBMA}$	23.9	47.5	69.1	86.0
[η]	30	34	44	73
Convers. (%)	26.3	18.5	16.0	15.7

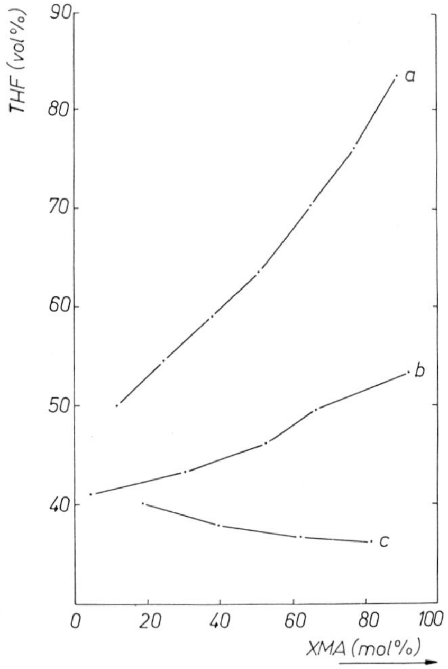

Fig. 9.11. Effect of the alcohol group in poly(methacrylate esters) on normal-phase elution characteristics of copolymers with styrene; (*a*): *stat*-copoly(styrene/methyl methacrylate), curve "a" from Fig. 6.9, (*b*): *stat*-copoly(styrene/ethyl methacrylate), (*c*): *stat*-copoly(styrene/*t*-butyl methacrylate). Silica columns, (a): 150 × 4.6 mm; $d_o = 6$ nm; $d_p = 5\ \mu$m; gradient *i*Oct/ (THF + 10% MeOH), see Fig. 9.3; (b) and (c): column 60 × 4 mm; $d_o = 5$ nm; $d_p = 5\ \mu$m; gradient: *i*Oct/THF (0–100% in 10 min). Sample data: S/MMA: Table 9.3, A–G; S/EMA: Table 9.5; S/TBMA: Table 9.6.1

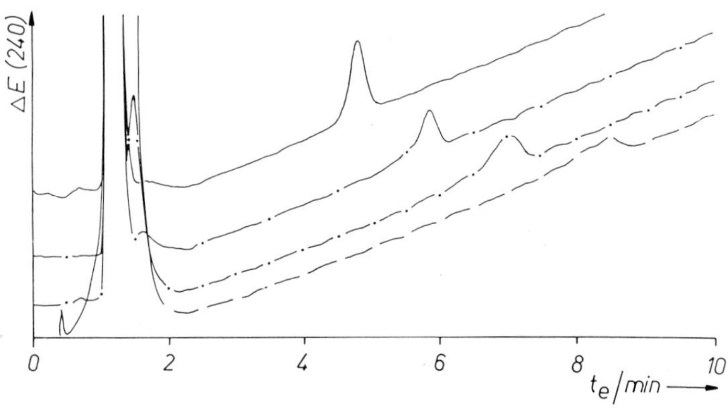

Fig. 9.12. Gradient elution of *stat*-copoly(styrene/*t*-butylmethacrylate) samples, D–A, from top to bottom) on phenyl bonded-phase column (60 × 4 mm; $d_o \leqq 5$ nm; $d_p = 7\ \mu$m), at 50 °C; gradient: methanol/tetrahydrofuran (10–70% in 12 min), flow rate 0.5 ml/min. Sample data: Table 9.6.1. UV signal at 259 nm. By courtesy of John Wiley & Sons, Ltd. [34]

decreases in the sequence MMA > EMA > TBMA, which can be understood on base of the solvent displacement theory [31] and as the consequence of increasing shielding of the ester linkage by alcohol residues. With the bulky TBMA group, the shielding is almost perfect and any remaining interactions between the polar surface of the silica and the copolymer are due to π-electrons of the styrene units.

Thus, a phenyl bonded-phase column was used in combination with a gradient MeOH/THF whose polarity decreased. Figure 9.12 shows the elution

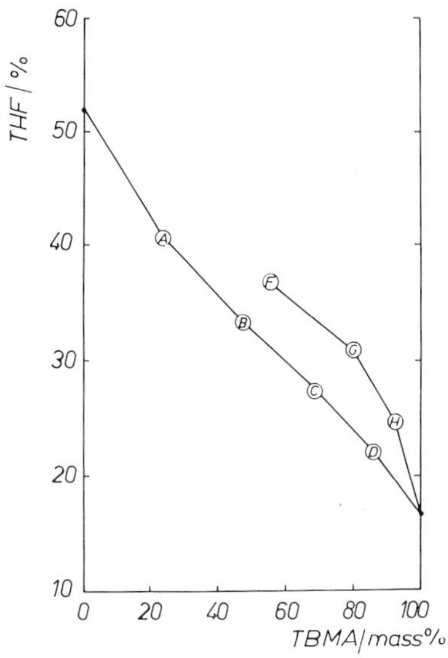

Fig. 9.13. Elution characteristics of copolymers from styrene and *t*-butyl methacrylate of either statistical (samples A-D) or block architecture (F-H). Column and gradient as in Fig. 9.12. Sample data: Tables 9.6.1 and 9.6.2

patterns obtained in a system like that. The copolymers elute in a sequence which reflects decreasing content in TBMA. The latest eluting specimen is sample A, which had been polymerized to about 26% conversion from a monomer mixture rather far apart from the azeotropic composition. The elution curve of sample A extends towards the position of sample C whose composition is close to the azeotropic one. Although the elution conditions still need optimization, the peaks are better graded than on a silica column and the chromatographic resolution is improved although column dimensions, flow rate, and gradient rate (5% THF/min) remained unchanged. The particle diameter of the phenyl bonded-phase packing was even larger ($d_P = 7\ \mu m$) than that of the silica. Hence, the improvement is really due to interactions between solute groups and column packing rather than to the elimination of unfavourable column kinetics.

In Fig. 9.13, the elution characteristics of *stat*-S/TBMA are displayed together with those measured with S/TBMA block copolymers. Note the similarity between the curve of the statistical copolymers and the characteristics (shown in Fig. 6.15) of the RP elution of S/EMA copolymers. Note also that the elution order of S/TBMA copolymers was not altered by the reversal of chromatographic conditions.

9.6.2 Block Copolymers

The four block copolymers listed in Table 9.6.2 were prepared by anionic polymerisation at $-78\ ^\circ C$ in a break-seal apparatus with cumylpotassium as an initiator [32, 33]. Due to partial chain termination of the living polystyrene at the addition of TBMA, some PS homopolymer was present in the samples F and

Table 9.6.2. Block copolymers, S/TBMA (M_s: 104.144, M_{TBMA}: 142.192)

Code	E	F	G	H
$100 w_{TBMA}$	26	55	80	92
$10^{-3} M^a$	300	240	92	112
Concomitant:	PS	PS	P(TBMA-S-TBMA)	
content (%)	40	15	40	
$10^{-3} M^a$	230	130	165	

[a] SEC (weight average)

E. Sample G contained approx. 40% of an TBMA-S-TBMA triblock copolymer of double MW. This product was formed by a side reaction of cumylpotassium yielding bifunctional α-methyl styrylpotassium. The concomitants were detected by SEC with RI and UV detection at 230 and 260 nm.

Figure 9.14 shows the elution patterns of the samples on a silica column under NP conditions. As with statistical S/TBMAs, retention increased with increasing content in styrene. The portions of samples E and F, which eluted in the period 10.9–11.3 min, indicate PS. In order to gain higher resolution (according to experience gathered with statistical S/TBMAs), the samples were also measured on a phenyl bonded-phase column with a RP gradient [34]. The gradation of the peak positions was indeed improved but the pattern of sample F changed in an unexpected manner: an additional peak appeared at lower elution volume, see Fig. 9.15.

With a C18 column and the same gradient, the peaks of the block copolymers were narrower, but sample F yielded a similar multimodal pattern with preliminary elution of part of the specimen.

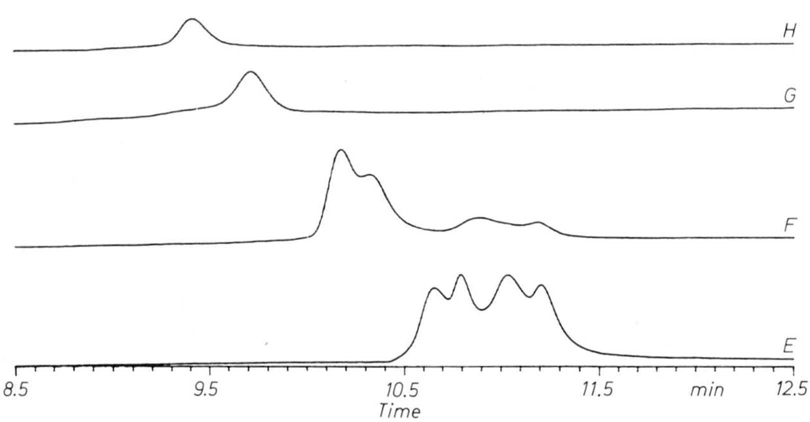

Fig. 9.14. Gradient elution of *Block*-copoly(styrene/*t*-butyl methacrylate) samples on silica by *iso*octane/tetrahydrofuran (0–100% in 20 min). Column: 60 × 4 mm; $d_o = 5$ nm $d_p = 5$ μm; 50 °C, flow rate 0.5 ml/min. Sample data: Table 9.6.2; $m_o = 8$ μg, $V_o = 20$ μl. By courtesy of John Wiley & Sons, Ltd. [34]

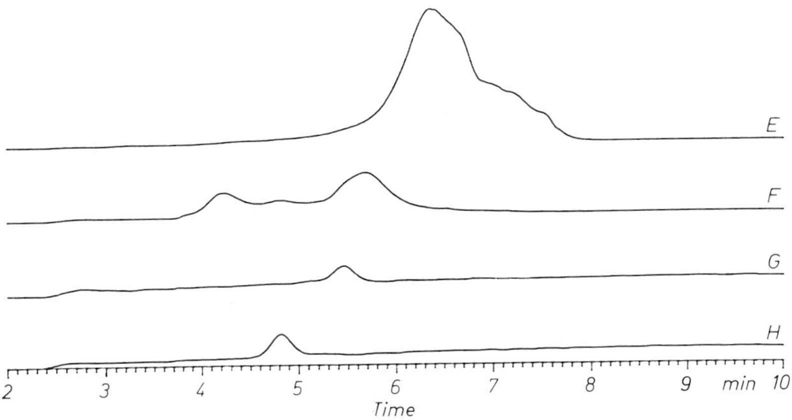

Fig. 9.15. Gradient elution of *block*-copoly(styrene/*t*-butyl methacrylate) on phenyl bonded-phase by methanol/tetrahydrofuran (0–100% in 10 min). Column: 60 × 4 mm; $d_o \leqq 5$ nm; $d_p = 7$ μm; 50 °C, flow rate 0.5 ml/min; samples as in Fig. 9.14, $m_o = 8$ μg, $V_o = 20$ μl. By courtesy of John Wiley & Sons, Ltd. [34]

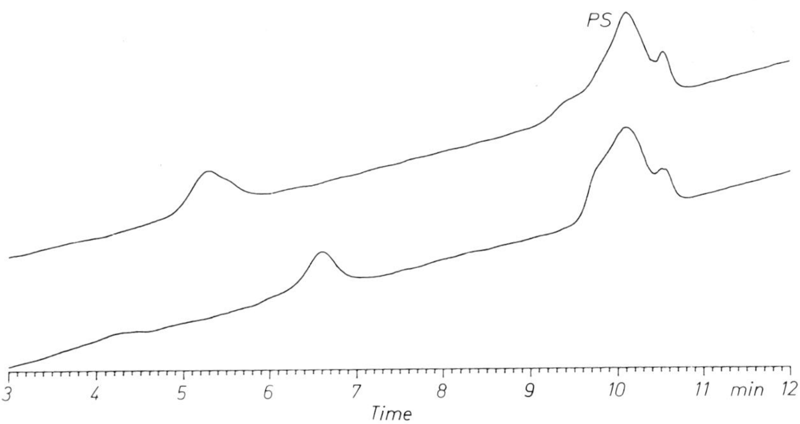

Fig. 9.16. Separation of admixed polystyrene 110 from S/TBMA block copolymer H (*upper curve*) or G (*lower curve*) on phenyl bonded-phase by methanol/tetrahydrofuran (10–70% in 12 min). Column: 60 × 4 mm; $d_o \leqq 5$ nm; $d_p = 7$ μm: 50 °C, flow rate 0.5 ml/min. By courtesy of John Wiley & Sons Ltd. [34]

In order to clarify whether or not the selectivity of the combination of a phenyl column with a gradient MeOH/THF would in principle suffice for separating S/TBMA block copolymers from admixed parent homopolymer, PS was investigated together with either the block copolymer G or H. (Both block copolymers showed narrow bands in gradient HPLC, and their mixtures with polystyrene could be adjusted to give an average TBMA content corresponding to that of sample F.) Figure 9.16 shows the elution of PS 110 (MW = 110,000) at about 10 min and that of the block copolymers H and G at 5.2 or 6.5 min,

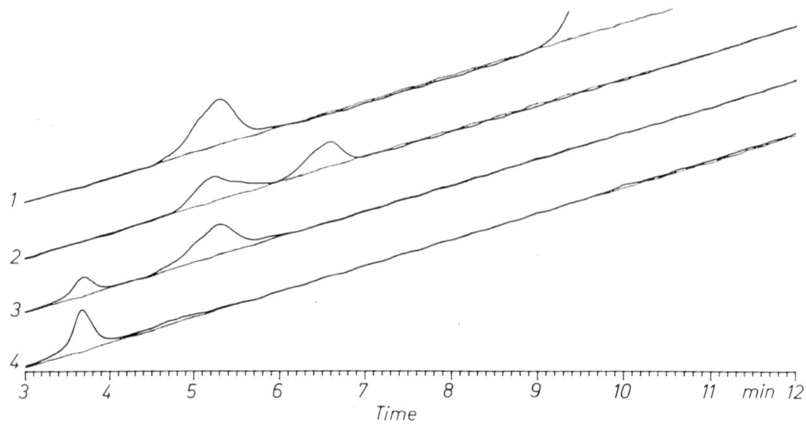

Fig. 9.17. Separation of S/TBMA block copolymers on phenyl bonded-phase by methanol/tetrahydrofuran (10–70% in 12 min). Column: 60 × 4 mm; $d_o \leqq 5$ nm; $d_p = 7$ μm; 50 °C, flow rate 0.5 ml/min. Curve *1*: block copolymer H mixed with PS 110 (5:1). Curve *2*: mixture of block copolymers G and H (1:1). Curve *3*: block copolymer H mixed with *t*-butyl methacrylate homopolymer (1:1). Curve *4*: *t*-butyl methacrylate homopolymer. By courtesy of John Wiley & Sons, Ltd. [34]

respectively. There is obviously no mutual disturbance on retention and elution in these mixtures of block products and homopolystyrene.

One can object that this was an easy task since the gap in composition was large. Therefore, attempts were made to separate also TBMA homopolymer from the samples G and H. Figure 9.17 shows the result. The uppermost trace once more shows the separation of the mixture of sample H (eluting between 4.8 and 5.6 min) and PS (eluting from 9.2 min onwards). Chromatogram No. 2 in Fig. 9.17 is from the mixture of the two block copolymers G and H. Although the difference in composition is only 12%, both samples were regularly eluted and baseline-separated. The third pattern was ontained on injection of the mixture of sample H (92% TBMA) and TBMA homopolymer. The lowest traces is from this parent polymer without admixed copolymer.

The usefulness of gradient HPLC for the investigation of block copolymers has been recognized also in the work of other authors. The clear separations obtained in favourable cases will certainly widen our insight into the structure of block copolymers and their behaviour in solution.

9.7 Alkyl Methacrylates and Acrylates with Different Side Groups

The effect of the alcoxy group in poly(methacrylate esters), which has been discussed in connection with Fig. 9.11, provides the chance of separating different methacrylate homopolymers through NP gradient HPLC. Figure 9.18 shows examples obtained by a gradient Tol/MEK on a silica column [17]. As expected

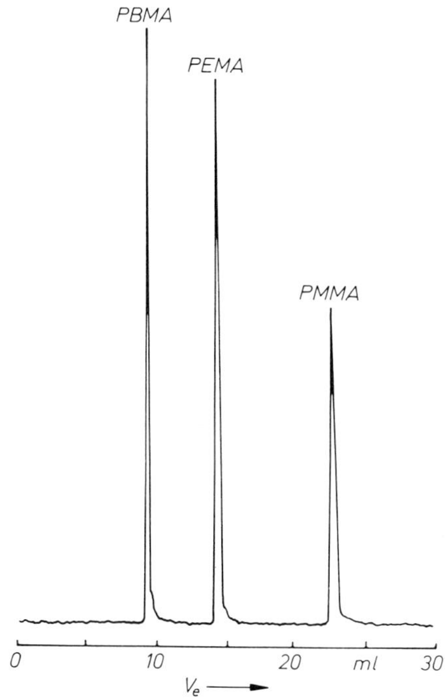

Fig. 9.18. Separation of polymethacrylate esters on silica by gradient toluene/2-butanone (MEK, 2–100% in 30 min). Column: 250 × 4.6 mm; $d_o = 6$ nm, $d_p = 5$ μm, flow rate 1 ml/min. Samples: poly(*n*-butyl-, ethyl-, and methyl methacrylate), $10^{-3} M_w$: 320, 340, or 93.3, respectively; $m_o = 10$ μg, $V_o = 10$ μl; monitored by evaporative light-scattering detector. From Ref. [17] with permission

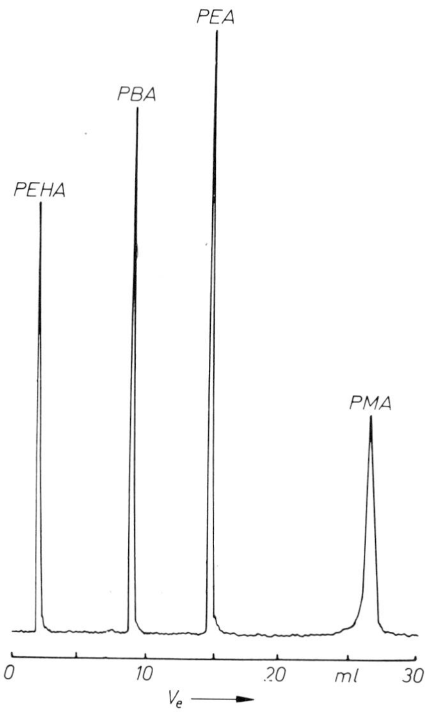

Fig. 9.19. Separation of polyacrylate esters (conditons see Fig. 9.18). Samples: poly(2-ethylhexyl-, *n*-butyl-, ethyl-, or methyl acrylate), $10^{-3} M_w$: 124, 119, 125, and 30.7, respectively. From Ref.[17] with permission

from the bulkiness of the side group. PBMA was less retained than PEMA or PMMA.

The same sequence holds for polyacrylate esters, see Fig. 9.19. In this case, poly(2-ethylhexyl acrylate), (PEHA), was an additional component of the injected sample. This polymer was eluted almost without retention in an elution volume only slightly larger than the interstitial volume of the column. The experimental conditions were identical in Figs. 9.18 and 9.19.

The relation between the size of an alcoxy group and its effect on retention can be understood on base of the solvent displacement theory [31]. The larger the size of a (weakly adsorbing) alkyl chain is, the more solvent molecules must be displaced from the stationary phase by the adsorption of a carboxyl group. But since a methylene unit has a larger effect if inserted into an alkyl chain (PMA–PEA: $\Delta t = 11.77$ min) than added in α-position (PMA–PMMA: $\Delta t = 4.2$ min), the adsorption of a carboxyl group obviously is additionally influenced by steric hindrance through a bulky alcoxy group. The gradation in retention time is surprisingly well reflected by respective differences in coil dimensions [35, 36].

The rather low MW of the PMA sample (weight average 30,700) could, in principle, be the reason for the observed smaller retention of PMA in comparison with PMMA but copolymers from MA and MMA showed the same effect at a much higher level in MW (see Sect. 9.8 and Table 9.7). Thus, the difference in retention is certainly due to the α-methyl group.

The difference can be caused by the solvent-displacing effect of a α-methyl, by reduced chain flexibility of PMMA which (in comparison with PMA) reduces the number of contacts with the stationary phase, or by the steric impairment of the approach of carboxyl groups towards the surface.

If the alcoxy chain is ethyl or longer, it fills so much space that an α-methyl cannot come close to the surface of an adsorbent. Thus, the "exposed" surface per repeat unit of PEMA is similar to that for PEA, for PBMA similar to that for PBA, etc. For instance, the bulkiness of *n*-butyl groups determines the conformation of acrylate polymers to such an extent that an α-methyl has no additional stiffening effect [35, 36]. This way, it can be understood why the differences in retention are large for methyl polyesters but small for the higher homoloques.

Decreasing retention with increasing size of alcoxy groups, as discussed here and with respect to Fig. 9.11, was also observed in TCM/EtOH eluent on a silica column [27].

9.8 Copolymers from Methyl Methacrylate and Methyl Acrylate

This system (*stat*(copoly(MMA/MA); $r_{MMA} = 0.3$, $r_{MA} = 1.5$ [37]) has no azeotropic point. Monomeric batches of any composition yield copolymers with a MA content above that of the starting mixture. In a first approximation, these copolymers can be looked at as consisting of MA microblocks separated by isolated MMA units.

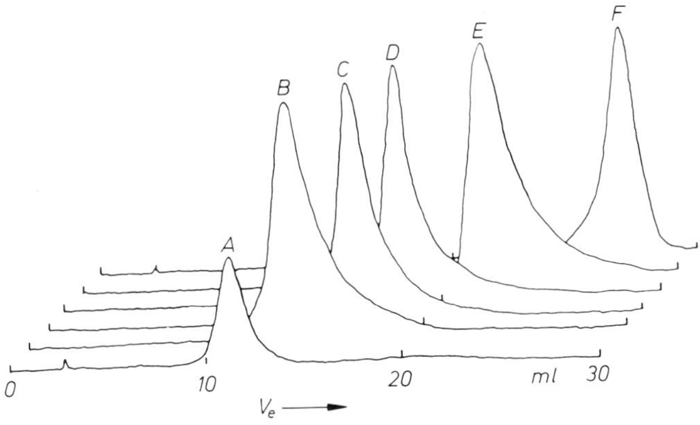

Fig. 9.20. Gradient elution of *stat*-copoly(methyl methacrylate/methyl acrylate) on silica by toluene/2-butanone (50–75% in 25 min). Column: 250 × 4.6 mm; $d_o = 6$ nm, $d_p = 5\ \mu$m; flow rate 1 ml/min. Peaks denoted by sample code, see Table 9.7 ("A": PMMA, "F": PMA). Elugrams monitored by evaporative light-scattering detector. From Ref. [17] with permission

The difference in retention of PMMA and PMA (see Figs. 9.18 and 9.19) in NP gradient HPLC offers the chance of separating copoly(MMA/MA) by composition. The specimens were prepared by free-radical polymerization in bulk with 0.2 mol% azo-bisisobutyronitrile initiator to less than 10% conversion. They were isolated by pouring into methanol. Characterization was limited to MW estimation from SEC curves (calibrated with PS standards). These data, together with the composition ratio of the starting mixtures and the calculated composition of copolymers at zero conversion, are given in Table 9.7.

The chromatograms measured on a silica column by a gradient Tol/MEK are shown in Fig. 9.20. The span in retention time between both parent homopolymers (samples A and F) was about 15 min or, at a gradient rate of 1%/min, 15% MEK. This corresponds to the difference in eluent composition for PMMA and PMA in Figs. 9.19 and 9.20 where a higher gradient rate was used.

A closer look at Fig. 9.20 shows that samples A–D, which were prepared from mixtures containing 0, 20, 40, or 60% MA, eluted in the intervall from 11.5 to 17 min, i.e. with a change in retention time of 5.5 min related to 60% difference in composition. The next copolymer (sample E), which was prepared from a mixture containing 80% MA, was retained longer than sample D by 3.4 min. The increase

Table 9.7. MMA/MA copolymers (M_{MMA}: 100.114, M_{MA}: 86.088)

Code	A (PMMA)	B	C	D	E	F (PMA)
$100 X_{MA}$[a]	0	20	40	60	80	100
$100\, w_{MA}$	0	35.0	54.3	70.0	84.8	100
$10^{-3} M$[b]	93.3	267.6	305.4	417.5	471	30.7

[a]monomer batch; [b]SEC (weight average)

in retention on transition to MA homopolymer, sample F, was even higher (about 6 min).

This deviation from linearity could not be caused by the *monomer reactivity ratios* of the system because these parameters predict, on 20% difference in feed composition, a larger change in copolymer composition for samples poor in MA and a smaller one for those rich in MA. This nonlinearity is opposite to the nonlinearity observed in retention behaviour. Especially the large influence on retention of a few MMA units is remarkable. It can be understood as an effect of solvent displacement or hindered approach towards the surface by MMA units, or a block-length effect (see Fig. 9.13) of MA sequences.

Since aliphatic polyacrylates and polymethacrylates show only weak UV signals in the wavelength regions suitable for detection, the chromatograms in Fig. 9.20 (as well as those in Figs. 9.18 and 9.19) have been monitored by using an evaporative light-scattering detector [17] (see Sect. 7.2.2) which enabled eluents with UV absorption to be used as gradient components (toluene and MEK, in the present case).

9.9 Copolymers from Styrene and 2-Methoxyethyl Methacrylate

Azeotropic copolymers of this system (*stat*(copoly(S/MEMA); $r_s = 0.46$; $r_{MEMA} = 0.48$) contain 59.0 mass% MEMA or 49.0 mol% S units. These copolymers have already been subject to manyfold studies. The instantaneous heterogeneity of an azeotropic S/MEMA copolymer has been investigated [38] by classical cross-fractionation, see Sect. 1.2, Table 1.1. The conversion effect on chemical composition distribution has been studied [39] by following the average copolymer composition and the composition of the residual monomer mixture as a function of conversion during the copolymerization reaction, see Sect. 2.4.1.

The samples investigated by gradient HPLC are listed in Table 9.8. Retention increasing with MEMA content could be achieved under NP conditions with *i*Oct/THF gradients on a column packed with CN bonded-phase material. Figure 9.21 shows chromatograms obtained that way [40], which had been monitored at 259 nm. At this wavelength, the absorptivity of the samples strongly depends on styrene content. For this reason, a merged plot of the recordings from samples A–H requires different attenuation which, in turn, causes different

Table 9.8. S/MEMA copolymers (M_s: 104.144, M_{MEMA}: 144.166)

Code	A	B	C	D	E	F	G	H	I
$100\,w_{MEMA}$	13.4	25.9	38.0	49.0	53.2	62.4	71.2	79.7	87.4
$10^{-3}\,M^a$	88	96	180	137	197	173	163	164	306
Convers. (%)	3	20	3	29	20	24	26	29	6

[a] by light scattering

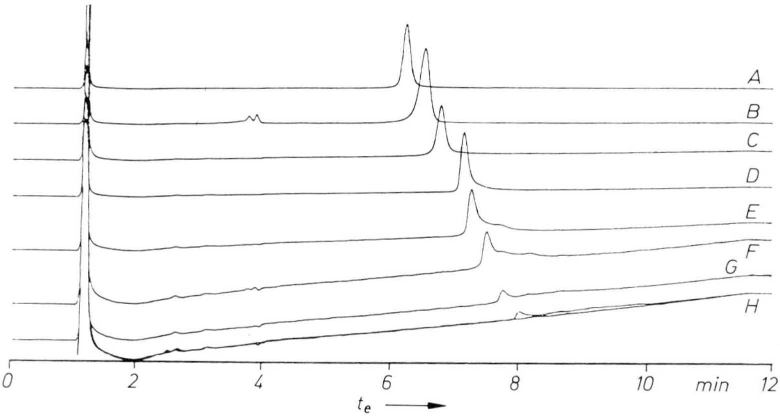

Fig. 9.21. Gradient elution of *stat*-copoly(styrene/2-methoxyethyl methacrylate) on CN bonded phase through *iso*-octane/tetrahydrofuran (0–100% in 10 min). Column: 60 × 4 mm, $d_0 \leqq 5$ nm, $d_p = 5\,\mu$m; samples A-H (see Table 9.8) injection of 10 μg each in THF, $V_0 = 10\,\mu$l. UV signal at 259 nm, attenuation graded according to signal size, *A*: 600 mAU full scale, *B*: 500, *C*, *D*: 300, *E*: 150, *F-H*: 75. By courtesy of John Wiley & Sons, Ltd. [40]

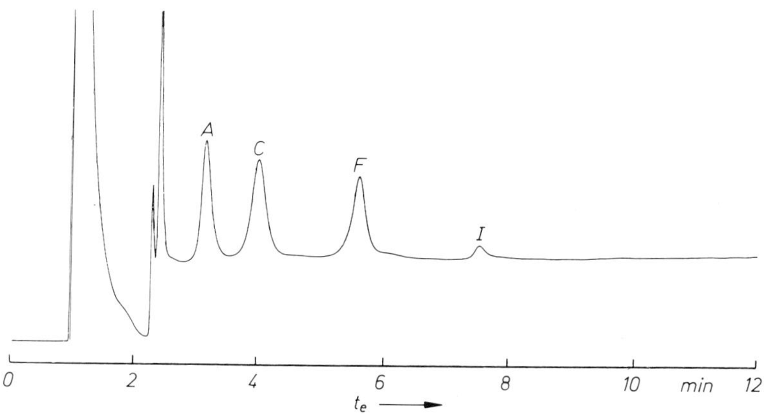

Fig. 9.22. Separation of the mixture of four *stat*-copoly(styrene/2-methoxyethyl methacrylate) samples on silica through a gradient *iso*-octane/methanol (MeOH increasing by 5%/min from 2% at t = 0 after a sudden transition in tetrahydrofuran concentration from zero to 35%). Column: 60 × 4 mm, $d_0 = 5$ nm, $d_P = 5\,\mu$m; 50 °C, flow rate 0.5 ml/min. Samples A, C, F, and I: see Table 9.8; injection of 20 μl sample solution in THF, containing 8.2 μg polymer in total. UV signal at 230 nm, 200 mAU full scale. By courtesy of John Wiley & Sons Ltd. [40]

slope factors of the gradient baseline. In spite of all efforts, the signal of sample H (79.7 mass% MEMA) is so poor that reasonable separations seemed rather unlikely under these chromatographic conditions. This pessimistic view is additionally based upon the tailing of the peaks which is noticable with samples containing 49% (sample D) or more MEMA and which is extreme with sample H.

Tailing and small signal size caused suspicion that THF was possibly not powerful enough for proper elution of samples rich in MEMA. This idea and the composition-dependence of a 259-nm signal gave incentive to modify the NP eluent system. As described in Sect. 5.6, a methanol gradient applied after a certain admixture of THF to the starting eluent (*i*Oct + 2% MeOH) improved the chromatograms definitely. This can be seen in Fig. 9.22 which is the chromatogram of a mixture of samples A, C, F, and I (the latter could not at all be seen under conditions as used in Fig. 9.21).

9.10 Copolymers of Styrene and Butadiene

This system (*stat*(copoly(S/Bd)); $r_S = 0.54$; $r_{Bd} = 1.57$) has no azeotropic point. Monomeric batches of any composition yield copolymers with a butadiene content above that of the starting mixture.

The samples listed in Table 9.9 were prepared in bulk with benzoyl peroxide as an initiator [41]. Gradient HPLC of these samples was performed on a column packed with beads of a highly crosslinked, porous copolymer of acrylonitrile and ethylene dimethacrylate (2:1) which had been obtained by suspension copolymerization in the presence of 3-methyl-1-butanol.

The mixture of all six samples A–F could be separated by composition through an exponential gradient Hex/TCM, see Fig. 9.23. Fractions obtained by SEC from sample D yielded HPLC peaks whose width increased with decreasing

Table 9.9. S/Bd copolymers (M_s: 104.144, M_{Bd}: 54.088)

Code	A	B	C	D	E	F
100 w_{Bd}	10.2	19.6	42.8	51.3	66.1	80.8
10^{-3} M^a	304	248	170	133	112	85
Convers. (%)	5.0	3.4	3.0	5.3	4.6	3.5

[a] SEC (weight average)

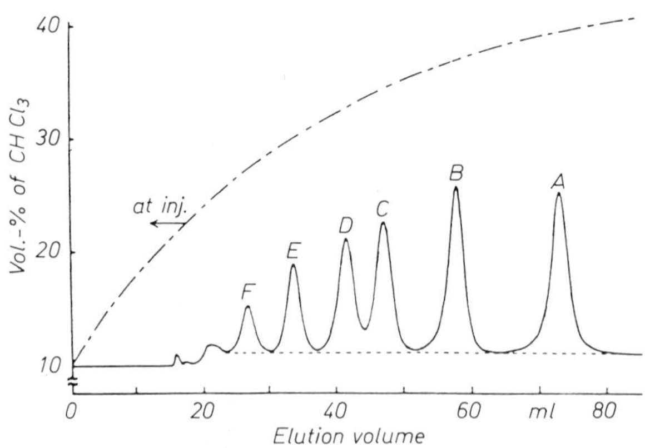

Fig. 9.23. Separation of six *stat*-copoly(styrene/butadiene) specimens on beads of cross-linked PAN ($d_P = 2.5$–$5\,\mu m$) by gradient *n*-hexane/trichloromethane (10–45%, as indicated). Column: 500 × 7.5 mm; peaks denoted by sample code, see Table 9.9. From Ref. [41] with permission

MW [41], which is consistent with the prediction of Stockmayer's theory [42]. The method was applied to common styrene/butadiene rubber (SBR) and to cold rubber [43]. The CCD of the latter was found to be narrower (peak width at 50% peak height: 4.5%) than that of the former (6.2%).

The method was also utilized for the investigation of two series of commercial triblock samples copoly(S–*b*–Bd–*b*–S) differing in styrene content and PS block length. As with the statistical samples shown in Fig. 9.23, retention increased with the overall styrene content. In addition, retention also increased with the sequence length of the S blocks. In other words, block copolymers were retained longer than statistical samples of the same composition. This behaviour is analogous to the results mentioned in Sects. 9.6 and 9.11.

9.11 Copolymers of Decyl Methacrylate and Methyl Methacrylate

The samples investigated were polymerized by group transfer (GT) copolymerization [44–46] with (1-methoxy-2-methyl-1-propenyl-oxy)trimethylsilane (MTS) as an initiator and tris(dimethylamino)sulfonium trimethylsilyldifluoride as a catalyst ($r_{MMA} = 1.66$, $r_{DMA} = 0.48$ [47]). The reaction is a living polymerization and was performed in THF at 0 °C under a blanket of pure nitrogen. The molar ratio of initiator to catalyst was chosen 50:1 and, in the preparation of block copolymers, the first monomer slowly added to the stirred solution of catalyst and initiator. When the reaction was complete after 1–2 h, another portion of catalyst was injected and the second monomer added. After completion, the living chain ends were terminated by ethanol. Statistical copolymers of high conversion were prepared in a corresponding manner but by the addition of a mixture of monomers. Because of the living reaction, even high-conversion copolymers have a narrow (intermolecular) CCD, similar to those of low-conversion samples of the system, which were prepared by the addition of catalyst to the solution of MTS initiator and monomer mixture in THF and quenching the reaction by EtOH after 30–120 s.

9.11.1 Statistical Copolymers

The samples listed in Table 9.10.1 were eluted on a CN bonded-phase column by a gradient *i*Oct/THF with concentration of THF increasing by 10%/min after an isocratic period of 1 min at a 99:1 composition [47]. The chromatograms

Table 9.10.1. Statistical copolymers, DMA/MMA (M_{DMA}: 226.348, M_{MMA}: 100.114)

Code	A(PDMA)	B	C	D	E	F(PMMA)
$100\,x_{MMA}$	0	25	40	50	75	100
$10\,w_{MMA}$	0	12.8	22.8	30.7	57.0	100
$10^{-3}\,M^{a}$	105	110	108	116	116	113

[a] SEC (weight average)

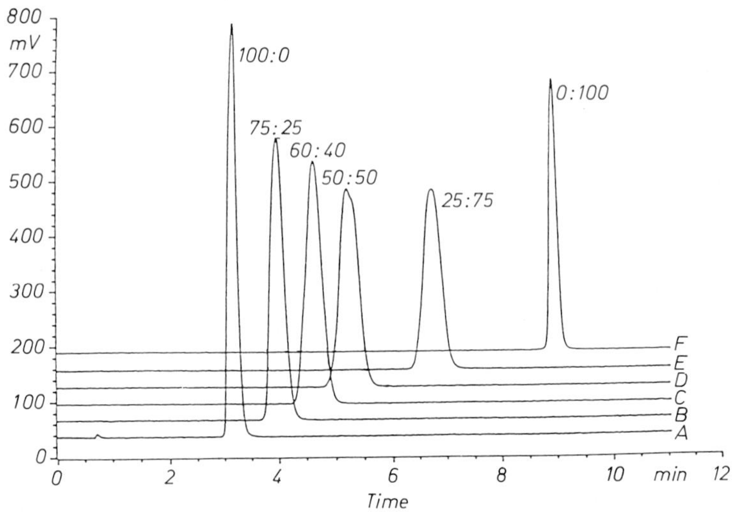

Fig. 9.24. Gradient elution of statistical copolymers of decyl methacrylate and methyl methacrylate on CN bonded-phase by *iso*-octane/tetrahydrofuran (1–81% in 8 min, start 1 min after injection). Column: 60 × 4 mm; $d_0 \leqq 5$ nm, $d_p = 5\,\mu$m; 40 °C, flow rate 1 ml/min; $m_0 = 25\,\mu$g (solutions in THF), $V_0 = 5\,\mu$l, samples see Table 9.10.1. Chromatograms monitored by an evaporative light-scattering detector. From Ref. [47] with permission

presented in Fig. 9.24 show retention increasing with MMA content of the copolymers.

9.11.2 Block Copolymers

Figure 9.25 is the analogous presentation of tracings obtained with block copolymers (data see Table 9.10.2) under identical conditions [47]. As in Fig. 9.24, retention increased with MMA content of the specimens but the composition effect is more pronounced than with statistical copolymers. This is another demonstration of the fact that a block copolymer is longer retained than a statistical copolymer of the same composition. A closer look at Fig. 9.25 shows that some poly(decyl methacrylate) homopolymer (PDMA) was concomitant to the block copolymers. (DMA was the starting monomer.) Figure 9.26 repeats the tracing of the 50:50 sample I from Fig. 9.25 for comparison with the chromatogram of another 50:50 copolymer (J in Table 9.10.2) produced by GT polymerization with MMA as a starting monomers. Here, some concomitant PMMA can be seen. Its elution time (8.75 min) was, due to its lower MW, shorter than that of the PMMA peak in Fig. 9.25 (8.9 min). The difference in MW between the concomitant and the main portion of sample J was large enough to reveal both components also in the SEC tracing, see Fig. 9.27a, but from the SEC curve no information about the nature of the extra component could be gained (besides its lower MW). Figure 9.27b presents the SEC recording of sample G which, according to the chromatogram *G* in Fig. 9.25, contained

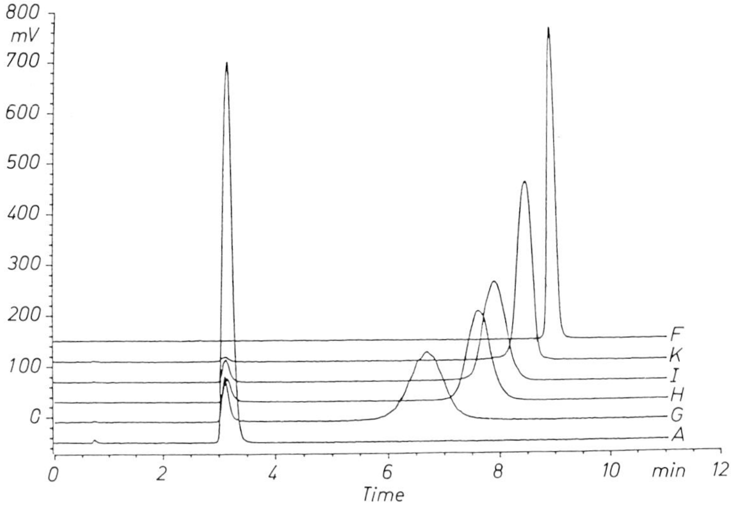

Fig. 9.25. Gradient elution of block copolymers of decyl methacrylate and methyl methacrylate (conditions as in Fig. 9.24). Samples see Table 9.10.2. From Ref. [47] with permission

Table 9.10.2. Block copolymers, poly(DMA-*b*-MMA) (M_{DMA}: 226.348, M_{MMA}: 100.114)

Code	G	H	I	J[a]	K
$100x_{MMA}$	25	40	50	50	75
$100\,w_{MMA}$	12.8	22.8	30.7	30.7	57.0
$10^{-3}\,M^{b}$	88	100	98	101	100

[a]poly (MMA-*b*-DMA) [b]SEC (weight average)

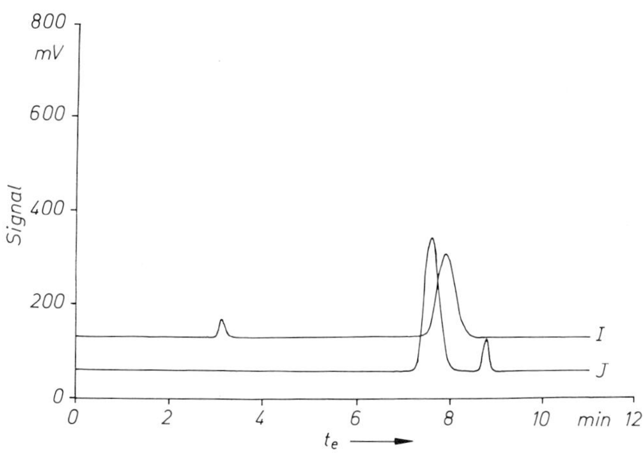

Fig. 9.26. Gradient elution of poly(methyl methacrylate-*block*-decyl methacrylate) 50:50 (sample J in Table 9.10.2) and of poly(decyl methacrylate-*block*-methyl methacrylate 50:50 (sample I in Table 9.10.2 and Fig. 9.25). Elution conditions see Fig. 9.24

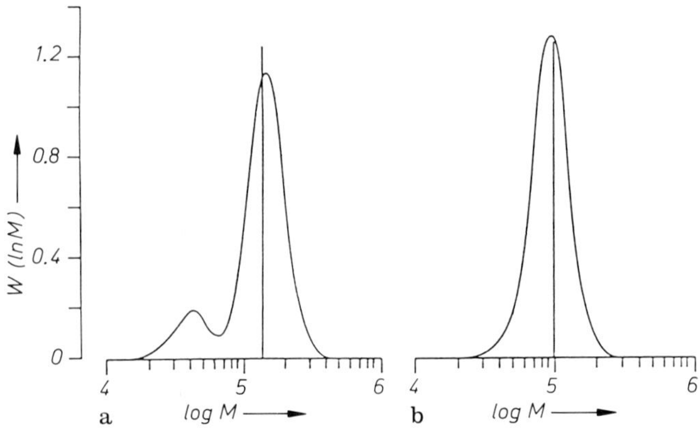

Fig. 9.27a,b. SEC tracings: (**a**) poly(methyl methacrylate-*block*-decyl methacrylate) 50:50 (sample J); (**b**): poly(decyl methacrylate-*block*-methyl methacrylate) 75:25 (sample G in Table 9.10.2, tracing of gradient elution: see Fig. 9.25). SEC on a bank of two columns (600 × 7.5 mm) packed with polystyrene gel (TSK GMH6, Toyo Soda, Japan), eluent THF, flow rate 1 ml/min, $V_0 = 300\,\mu l$, $m_0 = 300\,\mu g$, calibration by PMMA standards. From Ref. [47] with permission

some PDMA homopolymer. Here, the difference in MW was too small as to facilitate SEC separation of the species. Thus, judging a product only by its SEC curve can be misleading. This shows that progress in polymer characterization can indeed be achieved by gradient HPLC.

9.12 Graft Copolymers of Methyl Methacrylate onto Copoly(Ethylene/Propylene/Diene Monomer) (EPDM)

Graft copolymerizations usually yield mixtures containing the desired product, non-grafted precursor molecules, and homopolymers formed as a by-product during the grafting reaction. Figure 9.28 shows the SEC result of a polymer obtained by grafting MMA onto EPDM precursor [48]. The precursor had a MWD with an apex at about 250,000 and could be monitored by RI detector but not by UV at 239 nm, where MMA yielded a reasonable signal. PMMA by-product caused a large UV peak at $M = 2 - 3 \times 10^4$. The broad UV peak at $M = 3 - 6 \times 10^5$ is due to PMMA grafted onto EPDM. The SEC results enabled the amount of MMA to be estimated which was either homopolymerized or grafted, but SEC could not separate residual precursor from the desired graft product. This was possible by gradient HPLC in *i*Oct/THF on CN bonded-phase column. Figure 9.29 shows the signal monitored by an evaporative light-scattering detector. The first peak is from the non-grafted EPDM precursor which, under the chromatographic conditions employed, eluted unretained in a volume corresponding to the interstitial volume of the column. Peak No. 3 shows the desired graft product, whereas the more polar PMMA homopolymer was longer retained and eluted in peak No. 4. By number "2" a small peak is indicated which eluted with the solvent plug. The peak was caused by swept-through

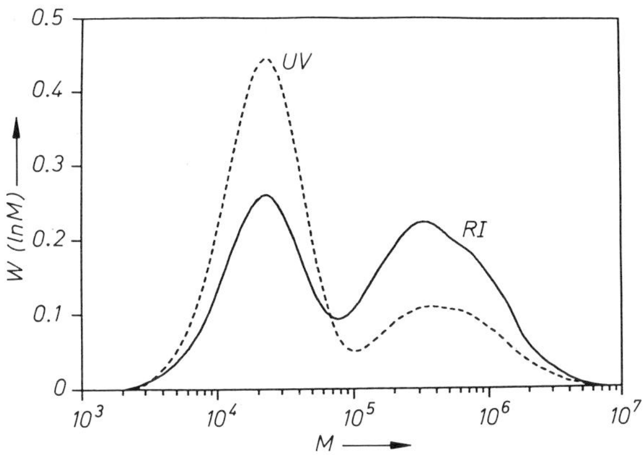

Fig. 9.28. Molecular-weight distribution of a graft product (PMMA grafted onto EPDM*). Components: PMMA (by-product, large UV peak at 20–30,000 MW), graft copolymer (UV peak at high MW), and EPDM precursor. SEC result, dual detection by differential refractometer and UV signal at 239 nm, $V_0 = 0.2$ ml, $m_0 = 0.2$ mg, eluent THF, flow rate 1 ml/min. From Ref. [48] with permission.*: copoly(ethylene/propylene/diene monomer)

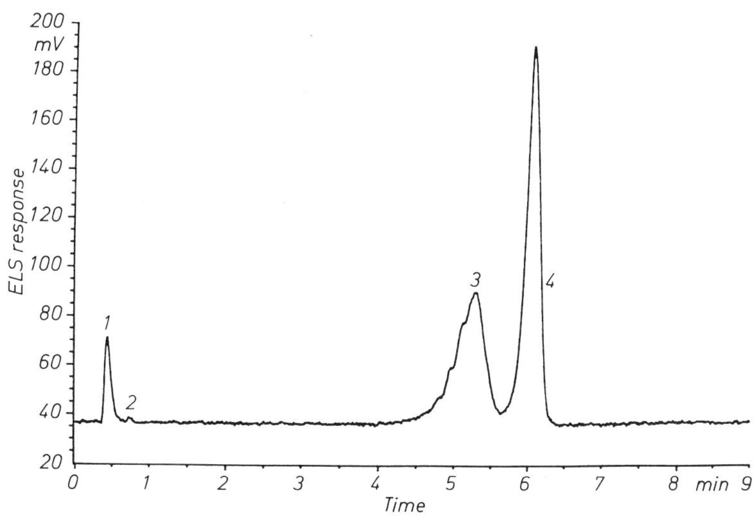

Fig. 9.29. Separation of EPDM-*graft*-PMMA (the same sample as in Fig. 9.28) on CN bonded phase by gradient *iso*-octane/tetrahydrofuran (1–100% in 6 min). Column: 60 × 4 mm; $d_0 \leqq 5$ nm; $d_P = 5\,\mu$m; 50 °C, flow rate 1 ml/min, elution monitored by evaporative light-scattering detector; $V_0 = 5\,\mu$l, $m_0 = 25\,\mu$g; peak identification: *1*: EPDM, *2*: solvent position, *3*: EPDM-*g*-MMA, *4*: PMMA. From Ref. [48] with permission

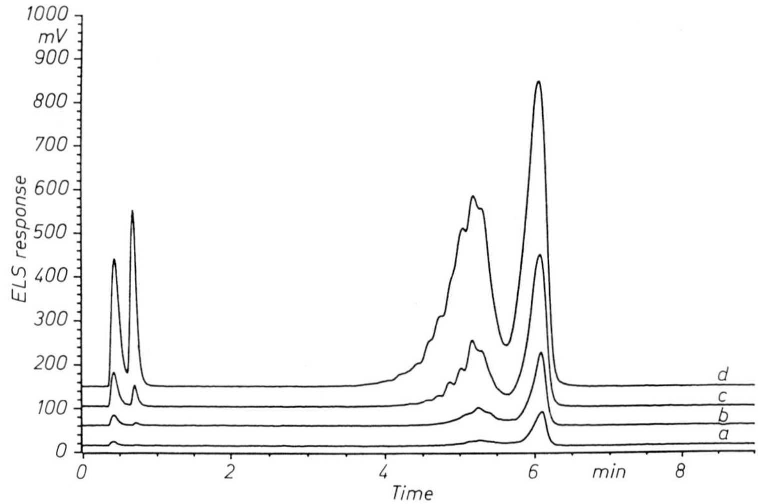

Fig. 9.30. Effect of sample size on swept-through portion of EPDM-*graft*-PMMA. Sample and conditions as in Fig. 9.29; injection volume *a*: 3 μl, *b*: 5 μl, *c*: 10 μl, *d*: 25 μl: From Ref. [48] with permission

polymer. The amount of this improperly retained portion progressively increased with sample size, see Fig. 9.30.

Chromatographic cross-fractionation of the reaction mixture is reported in Sect. 10.6.

9.13 Epoxy Composite Formulations, Prepolymers of Phenol-Formaldehyd Resins, Macromonomers, and Telechelic Prepolymers

Although of rather low MW, prepolymers are included in this survey because gradient HPLC is one of the most promising methods for the characterization of these materials. Variations in composition of epoxy formulations do not always show up in the mechanical performance of a test-piece but may have severe effects on the long-term performance of final products.

HPLC investigations of epoxy resins (EP) were performed in the NP mode [49, 50] as well as in the RP mode [50–56]. A survey has been provided by Mestan and Morris [57].

Since EP composite formulations usually contain fillers, the resin components must be extracted from the crude material. Three consecutive extractions with THF at room temperature proved successful. Filtering before injection is indispensable.

Excellent separations have been gained on RP C18 columns by gradients W/AcN, see Fig. 9.31 [54]. The eluent components allowed UV detection at 230 nm without disturbance. In both separations shown, the gradient was

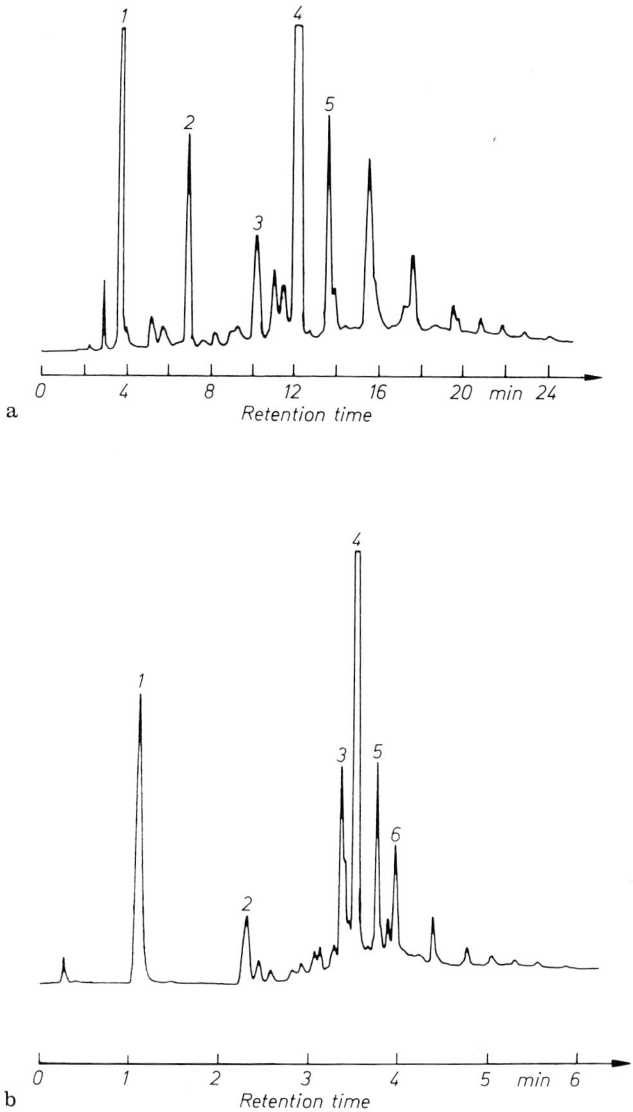

Fig. 9.31 a, b. Separation of a THF extract ($c \approx 6$ g/l) from epoxy composite formulation (Narmco Rigidite 5208/WC3000 prepreg) on RP C18 by gradient water/acetonitrile (AcN) at 25 °C, monitored at 230 nm. (**a**): Column 300 × 4 mm; $d_P = 5\ \mu m$; 50–100% AcN in 15 min, flow rate 1 ml/min; (**b**): Column 40 × 4.6 mm; $d_P = 3\ \mu m$; 20–100% AcN in 3 min, flow rate 2 ml/min. (*1*): 4, 4′diaminodiphenylsulfone (DDS, curing agent), (*2*): acetophenone (added as an internal standard), (*3*): initial reaction product of DDS and TGDDM (see peak #*4*), (*4*): tetraglycidyl-4,4′-diamino-diphenylmethane (TGDDM), (*5*): diglycidyl ether of bisphenol A, (*6*): not identified (probably higher reaction product of DDS and TGDDM). From Ref. [54] with permission

followed by isocratic elution at 100% AcN for 15 (Fig. 9.31a) or 2 min (9.31b). A 40-mm column with 3-μm packings yielded, at reduced analysis time, separations almost as good as a 300 mm column with 5-μm particles.

Prepolymers of phenol-formaldehyd resins contain numerous compounds differing in size and structure. They have been separated isocratically on polar [58] or nonpolar columns [59]. Gradient elution was performed with W/THF [60], W/THF (with addition of phosphoric acid) [61] or W/AcN (with addition of phosphoric acid) [62] on RP columns. W/MeOH gradients in combination with a RP C18 column were also employed [63, 64]. A gradient of the latter components (W/MeOH, 10–90% within 40 min, exponentially) proved successful also in combination with a microcolumn (150 × 0.7 mm) packed with RP C18 [65]. Recently, temperature-programming supercritical-fluid chromatography with decreasing column temperature was used for the separation of phenol-formaldehyd prepolymers. The eluent was liquefied carbon dioxide (300 μl/min flow rate) with a contineous addition of ethanol of either 100 or 50 μl/min flow rate [66]. Chromatograms were monitored by UV absorption at a suitable wavelength in the range between 210 [65] and 280 nm [59].

Macromonomers are oligomers with polymerizable end groups. Compounds of that kind prepared from aryloxy(polyethyleneoxy)alcohols and acrylic acid or vinylbenzyl chloride have been investigated by HPLC, SEC, and gas chromatography [67]. The aryl compound in the alcohol component was either phenyl or nonylphenyl. Of the chromatographic techniques investigated, RP gradient HPLC with W/AcN on a C18 column was found to be the most useful method for macromonomer analysis. It could be applied over a wide range of MW and had favourable selectivity towards the macromonomer components. For instance, reversed-phase HPLC of a nonyl-phenoxy poly(ethyleneoxy)acrylate sample ("nonyl-PO-P(EO)acrylate") at nominal ten EO repeat units yielded separation of a PO-P(EO) acrylate admixture from the sequence of nonyl-PO-P(EO)alcohols ("AL", retention times between 10 and 23 min) and of nonyl-PO-P(EO)acrylates ("AC", 23–33 min), see Fig. 9.32. In both sequences, retention increased with decreasing length of the ethylene-oxide block. With the "AL" sequence, individual peaks could be detected for

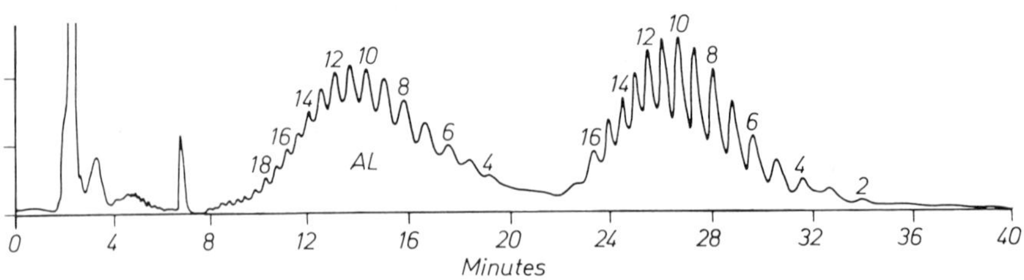

Fig. 9.32. Separation of the reaction mixture of nonylphenoxy poly(ethylenoxy) alcohol (*AL*) and acrylic acid by RP gradient HPLC on a C18 column (250 × 4.6 mm; $d_0 = 5$ nm, $d_P = 3$ μm); gradient water/acetonitrile (60–80% in 16 min), flow rate 1 ml/min, $m_0 = 40$ μg, $V_0 = 40$ μl; UV detection at 215 nm. (The figures at the peaks denote the number of ethyleneoxy units of the respective molecules. The precursor *AL* had nominal 10 repeat units). From Ref. [67] with permission

components having 4–18 repeat units, and with the "AC" sequence for components having 2–16 units. Peaks have been identified either by analysing individual oligomers or by comparison with chromatograms of samples with known oligomer distribution.

Telechelic prepolymers of diglycidylether-bisphenol A and aniline with either glycidyl [55] or aminoalcohol end groups [56] were chromatographed on RP C18 columns with gradients W/AcN. The RP HPLC was capable of separating the polymer homologues and, within each member of the series, the diastereomers.

9.14 Ethylene/α-Olefin Copolymers

Copolymers of ethylene and α-olefins ("linear low-density polyethylene", LLDPE) are widely used as films of excellent stretchability and stiffness. The α-olefin comonomers cause short chain branching (SCB). SCB (together with long chain branching) also makes the difference between high-density PE and common LDPE produced in high-pressure processes.

SCB diminishes the crystallizability and, thus, eventually lowers the dissolution temperature of the polymer. The reduction of dissolution temperature is, in turn, the basis for a method of separating olefin copolymers by composition which is known as temperature rising elution fractionation (TREF). The main goal of this method is the evaluation of SCB distribution [68].

The central part of a TREF apparatus is a thermostated column packed with an inert support material. The polymer to be investigated is dissolved in a suitable solvent at elevated temperature. The hot sample solution is introduced into a heated column. Then the flow is stopped and the temperature of the column slowly lowered. Thus, the sample is deposited onto the surface of the column packing. The subsequent separation is performed by controlled raising the temperature of the column and eluting fractions. A scheme of a TREF apparatus is shown in Fig. 9.33.

In analytical TREF, the concentration of the eluate is monitored as a function of elution temperature. The method has been developed by Wild and Ryle in 1977 [69]. The authors recognized that slow cooling rates improved the subsequent separation. Hence, the cooling rate was set to 1.5 K/h in routine analyses.

The time-consuming cooling step became the bottleneck of the technique. Thus, the slow crystallization procedure was performed simultaneously on a battery of exchangeable columns in a separate cooling bath. The columns could be easily inserted into the true TREF apparatus where elution was performed at a flow rate of about 6 ml/min and a heating rate of 8 K/h. Thus, a sample of 200 mg was fully eluted within about 8 hours. Reduction of sample amount to 50 mg [70, 71] or even 10 mg [72] and increase of heating rate to 20 K/h has been reported in subsequent papers dealing with TREF. Another promising version is fractionating crystallization from a sample solution in a test tube and loading the solid precipitate into a dismantled TREF column. The eluents used were xylene, 1,2,4-trichlorobenzene [69–71] or α-chloronaphthalene [72].

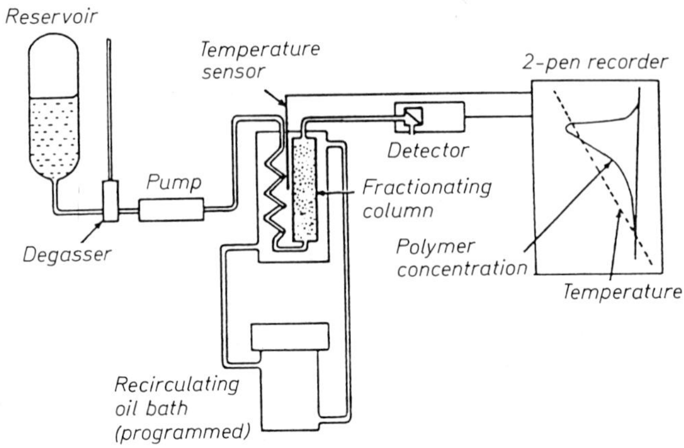

Fig. 9.33. Scheme of a TREF apparatus. From Ref. [70] with permission

In another device, the temperature was raised gradually after a rather fast cooling period. The automated apparatus required only 2 mg sample material and was run with *o*-dichloro benzene [73–75]

By means of an appropriate calibration relating the elution temperature to SCB density, the primary elugrams can be converted into differential SCB distribution curves. The calibration has been established in several investigations

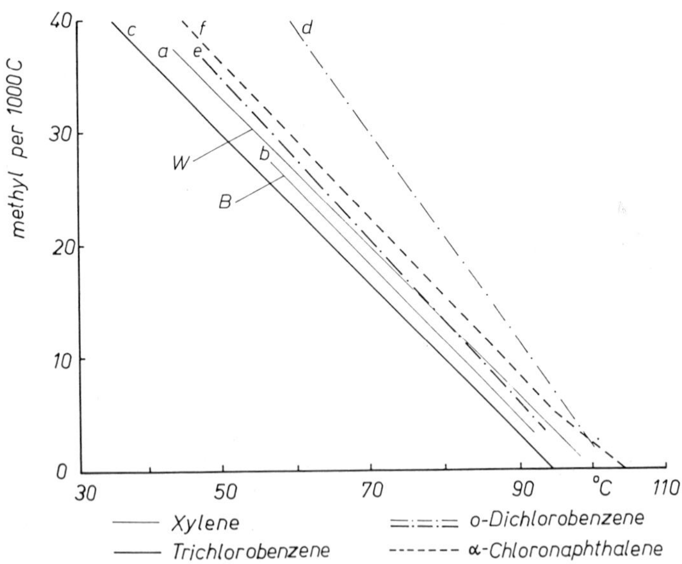

Fig. 9.34. Calibration of TREF elution temperature as a function of methyl group concentration determined by IR absorption. Results in different eluents, (xylene, 1, 2, 4-trichloro-benzene, *o*-dichloro benzene, α-chloronaphthalene, as indicated) (W: Wild et al. (1977) [69], B: Bergström et al. (1979) [77])

and shows the linear increase of elution temperature with decreasing number of $—CH_3$ groups, see Fig. 9.34. It is worthwhile to note that, in spite of different experimental conditions, the calibration lines are close together and have almost identical slope.

By TREF investigations it became clear that LLDPE is composed of an almost linear portion and of material with a broad SCB distribution, whereas common LDPE has a broad but unimodal distribution of SCB.

Recently, an automated apparatus has been described which is capable of analysing eight samples a day [76].

9.15 Copolymers of Styrene and Acrylic Acid

Statistical copolymers of styrene and acrylic acid were synthesized [78] by free radical polymerization in THF solution at 50 °C with 2, 2′-azo(bis-*iso*-butyronitrile) as an initiator. After about 10% conversion, the reaction mixtures were poured into *n*-heptane. The precipitated copolymers were dried at 50 °C under vacuum.

Reversed-phase HPLC was performed [78] on a column 100 × 4 mm (packed with LiChrosorb RP18) by a gradient water/THF (0–100% in 20 min) at 39.8 °C and a flow rate 0.8 ml/min. Proper retention and elution could be achieved in the whole range of copolymer composition. Retention increased with decreasing content in acrylic acid.

Normal-phase HPLC on silica was possible only with copolymers containing 0–30% acrylic acid. Samples with a higher content in the acid unit could not be eluted from the column.

9.16 Separation of PMMA Isomers

Poly(methyl methacrylate) with either syndiotactic or isotactic configuration could be separated under normal-phase and reversed-phase conditions [79]. Among the NP systems investigated was the combination of a column packed with crosslinked polyacrylonitrile and a gradient *n*-hexane/dichloromethane (40–100% in 25 min), among the RP systems a column packed with crosslinked polystyrene and a gradient nitromethane/DCM (5–30% in 25 min). Both combinations were capable of separating mixtures of *s*- and *i*-PMMA at a reasonable resolution. Under all NP conditions studied, *s*- PMMA was longer retained, under all RP conditions it was earlier eluted than *i*-PMMA. The inversion of elution order indicated that the separation was not due to the difference in molecular weight of the samples but to an apparently higher polarity of syndiotactic PMMA.

9.17 References

1. Scholtan W, Kwoll FJ (1972) Makromol Chem 151: 33
2. Teramachi S, Obata R, Yamashita K, Takemoto N (1977) J Macromol Sci A11: 535

3. Ogawa S, Sakai M (1981) J Polym Sci A-2, Polym Phys 19: 1377
4. Teramachi S, Fukao T (1974) Polym J 6: 532
5. Glöckner G, Francuskiewicz F, Müller K-D (1971) Plaste and Kautschuk 18: 654
6. Glöckner G, Kroschwitz H, Meissner C (1982) Acta Polymerica 33: 614
7. Glöckner G (1983) Pure Appl Chem 55: 1553
8. Glöckner G, van den Berg JHM (1984) Chromatographia 19: 55
9. Glöckner G, van den Berg JHM, Meijerink NL, Scholte TG, Koningsveld R (1984) Macromolecules 17: 962
10. Glöckner G, Koningsveld R (1983) Makromol Chem, Rapid Commun 4: 529
11. Glöckner G, van den Berg JHM, Meijerink NL, Scholte TG (1986) In: Kleintjens L and Lemstra P (eds) Integration of fundamental polymer science and technology, Elsevier, Barking, UK p 85
12. Glöckner G, van den Berg JHM, (1987) J Chromatogr 384: 135
13. Glöckner G, van den Berg JHM (1987) Chromatographia 24: 233
14. Danielewicz M, Kubin M, Vozka S (1982) J Appl Polym Sci 27: 3629
15. Teramachi S, Hasegawa A, Shima Y, Akatsuka M, Nakajima M (1979) Macromolecules 12: 992
16. Sato H, Takeuchi H, Tanaka Y (1986) Macromolecules 19: 2613
17. Mourey TH (1986) J Chromatogr 357: 101
18. Sparidans RW, Claessens HA, van Doremaele GHJ, van Herk AM (1990) J Chromatogr 508: 319
19. Danielewicz M, Kubin M (1981) J Appl Polym Sci 26: 951
20. Glöckner G, van den Berg JHM (1986) J Chromatogr 352: 511
21. Teramachi S (1984) personal communication
22. Mori S (1987) J Chromatogr 411: 355
23. Mori S, Uno Y (1987) Analyt Chem 59: 90
24. Teramachi S, Hasegawa A, Motoyama K (1987) Polym Preprints, Japan 36: E441, 3169
25. Sato H, Mitsutani K, Shimidzu I, Tanaka Y (1988) J Chromatogr 447: 387
26. Mori S (1989) J Appl Polym Sci 38: 95
27. Mori S, Mouri M (1989) Analyt Chem 61: 2171
28. Glöckner G, Stickler M, Wunderlich W (1987) Fresenius Z Anal Chem 328: 76
29. Glöckner G, Stickler M, Wunderlich W (1988) Fresenius Z Anal Chem 330: 46
30. Glöckner G (1987) J Chromatogr 403: 280
31. Snyder LR (1968) Principles of adsorption chromatography, Marcel Dekker, New York
32. Rahlwes D, Kirste RG (1977) Makromol Chem 178: 1793
33. Müller AHE (1981) Makromol Chem 182: 2863
34. Glöckner G, Müller AHE (1989) J Appl Polym Sci 38: 1761
35. Wunderlich W (1970) Angew Makromol Chem 11: 201
36. Wunderlich W (1972) J Polym Sci C Polym Symp 39: 145
37. Cranwels K, Smets G (1950) Bull Soc Chim Belges 59: 443
38. Stejskal J, Kratochvíl P, Straková D (1981) Macromolecules 14: 150
39. Stejskal J, Kratochvíl P, Straková D, Procházka O (1986) Macromolecules 19: 1575
40. Glöckner G, Stickler M, Wunderlich W (1989) J Appl Polym Sci 37: 3147
41. Sato H, Takeuchi H, Suzuki S, Tanaka Y (1984) Makromol Chem, Rapid Commun 5: 719
42. Stockmayer WH (1945) J Chem Phys 13: 199
43. Sato H, Takeuchi H, Tanaka Y (1985) Int Rubber Conf Kyoto [18B15]: 596
44. Hertler WR, Sogah DY, Farnham WB, Rajanbabu TV (1983) J Am Chem Soc 105: 5706
45. Mai P, Müller AHE (1987) Makromol Chem, Rapid Comm 8: 247
46. Müller MA, Stickler M (1987) Makromol Chem, Rapid Comm 7: 575
47. Müller MA, Augenstein M, Dumont E, Pennewiß H New Polymeric Materials (submitted for publ.)
48. Augenstein M, Stickler M (1990) Makromol Chemie 191: 415
49. Hagnauer GL, Setton I (1978) J Liq Chromatogr 1: 55
50. Hagnauer GL (1980) Polym Compos 1: 81
51. Crabtree DJ, Hewitt DB (1977) Chromatogr Sci 8: 63
52. Shiono S, Karino I, Ishamura A, Enomoto J (1980) J Chromatogr 193: 243
53. Byrne CA, Hagnauer GL, Schneider NS (1983) Polym Compos 4: 206
54. Noel D, Cole KC, Hechler J-J, Chouliotis A, Overbury KC (1986) J Appl Polym Sci 32: 3097
55. Klee J, Hörold H-H Tänzer W, Fedtke M (1986) Acta Polym 37: 272
56. Klee J, Hörold H-H, Tänzer W, Fedtke M (1987) Angew Makromol Chem 147: 71
57. Mestan SA, Morris CEM (1984) J Macromol Sci, Rev Macromol Chem Phys C24: 117
58. Šebenik A, Lapanje S (1978) Angew Makromol Chem 70: 59

59. Casiraghi G, Sartori G, Bigi F, Cornia M, Dradi E, Casnati G (1981) Makromol Chem 182: 2151
60. van der Maeden FPB, Biemond MEF, Janssen PCGM (1978) J Chromatogr 149: 539
61. Much H, Pasch H (1982) Acta Polym 33: 366
62. Werner W, Barber O, Chromatographia 15 (1982) 101
63. Mechin B, Hanton D, LeGoff J, Tanneur JP (1984) Europ Polym J 20: 333
64. Mechin B, Hanton D, LeGoff J, Tanneur JP (1986) Europ Polym J 22: 115
65. Preussler V, Slais K, Hanus J (1987) Angew Makromol Chem 150: 179
66. Mori S, Saito T, Takeuchi M (1989) J Chromatogr 478: 181
67. Schulz WW, Kaladas J, Schulz DN (1984) J Polym Sci A-1, Polym Chem 22: 3795
68. Glöckner G (1990) J Appl Polym Sci, Appl Polym Symp, Barth HG, ed. 45: 1
69. Wild L, Ryle TR (1977) Polym Prepr, Am Chem Soc, Polym Chem Div 18: 182
70. Wild L, Ryle TR, Knobeloch DC, Peat IR (1982) J Polym Sci A-1, Polym Phys 20: 441
71. Wild L, Ryle TR, Knobeloch DC (1982) Polym Prepr, Am Chem Soc, Polym Chem Div 23: 133
72. Kelusky EC, Elston CT, Murray RE (1987) Polym Engn Sci 27: 1562
73. Usami T, Gotoh Y, Takayama S (1986) Macromolecules 19: 2722
74. Nakano S, Goto Y (1981) J Appl Polym Sci 26: 4217
75. Gotoh Y, Usami T, Takayama S (1988) Poster Presentation at 1st ISPAC Meeting Toranto, June 2–3
76. Hazlitt L (1990) J Appl Polym Sci, Appl Polym Symp, Barth HG, ed. 45: 25
77. Bergström C, Avela E (1979) J Appl Polym Sci 23: 163
78. Sparidans RW (1990) Graduation research report, Eindhoven University of Technology, Instrumental Analysis and Polymer Chemistry
79 Sato H, Sasaki M, Ogino K (1989) Polym J 21: 965

10 Chromatographic Cross-Fractionation

10.1 Scope of this Chapter

Literally, the term chromatographic cross-fractionation refers to any combination of chromatographic methods capable of evaluating the distribution in size and composition of copolymers. The present pages are intended to give an overview on CCF methods which are susceptible of automation. Hence, these pages concentrate on CCF by combination of SEC and gradient HPLC, preferably in a sequence as illustrated in Fig. 10.1. (The opposite sequence has been also used [1, 2]. It has the advantage that a final *isocratic elution* can be carried out with multiple detection, thus providing additional information with reference to the composition of the fractions investigated.) Besides gradient HPLC, temperature rising elution fractionation (TREF) will be dealt with as another counterpart to SEC.

The chapter will not include detailed discussion of combinations of thin-layer chromatography (TLC) and SEC, which also have been applied to analysis according to MWD-CCD of copolymers. Among the remarkable investigations of that kind are analyses of a high-conversion sample of *stat*-copoly (S/MA) reported by Teramachi et al. [3], of branched polystyrenes by Belenkii and Gankina [4], of *block*-copoly (MMA/S/MMA) by Belenkii et al. [5], or of *stat*-copoly(S/EMA) by Tacx and German [6, 7]. These studies were all performed by SEC prefractionation and TLC-analysis of the fractions.

The inverse sequence with TLC-analogous prefractionation in a column packed with dry silica and SEC analysis of the fractions was used by Inagaki et al. [8] in a thorough study of the complex distribution of a *diblock*-copoly (S/MMA) sample. Figure 10.2 displays the SEC curves of the fractions whose styrene content varies between 21 and 84.9%.

The work is of especial importance because the authors also investigated the sample in traditional fashion by dual detection SEC. The result is shown in Fig. 10.3 and indicates an almost constant composition along the SEC elution curve. The reason for the failure of dual detection SEC can be found in Fig. 10.2: the SEC tracings of the fractions are similar in shape and position, thus an almost perfect balance of chemically different portions in dual-detection SEC gives the wrong impression of a negligible chemical heterogeneity of the whole sample. The work is a brilliant demonstration "indicating that no exact information of the composition heterogeneity can be drawn from GPC data", as the authors [9] stated.

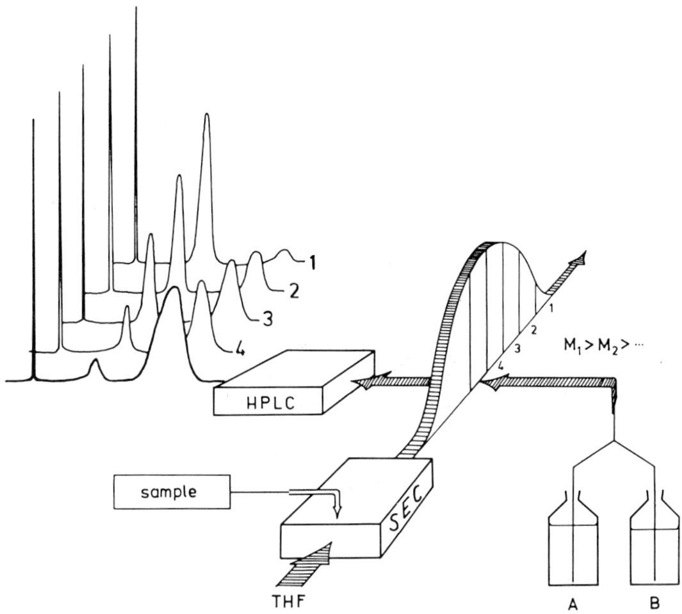

Fig. 10.1. Chromatographic cross-fractionation by SEC prefractionation and gradient HPLC of eluate fractions without additional treatment (schematic picture). HPLC tracings indicate (from left to right) a solvent peak, a peak of a high-MW copolymer which is poor in polar monomeric units, and a peak of a low-MW copolymer rich in polar units (normal-phase separation supposed)

Most of the publications mentioned have been discussed in more detail in a recent review on chromatographic cross-fractionation [10].

10.2 Comparison of Calculated and Measured Curves of Chemical Composition Distribution

In several publications, the chemical composition distribution (CCD) of copolymers has been evaluated by traditional fractionation and compared with the differential or integral CCD curves calculated from theory. There might be the idea of taking over this strategy to chromatographic separations by comparing calculated curves with gradient chromatograms of a whole sample (adequate calibration provided). Unfortunately, this straightforward approach will work only in favourable cases, i.e. in the investigation of samples whose distribution in MW is extremely narrow (e.g. SEC fractions, see Fig. 9.6) or under HPLC conditions which result in separation strictly according to composition without any influence of MW. In general there is some superimposition of MW and composition effects in gradient HPLC. Thus, a warning must be given against interpreting the HPLC curve of crude copolymers as the immediate picture of their differential CCD curve; this procedure can cause erroneous results.

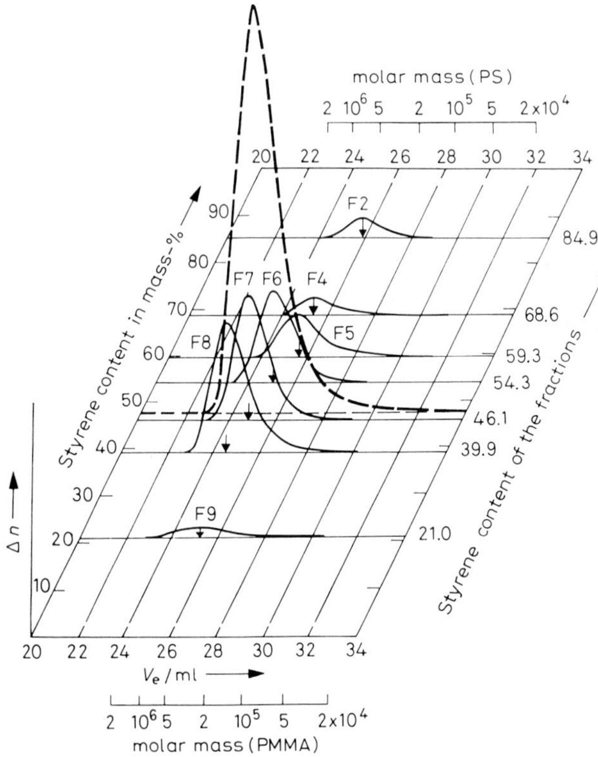

Fig. 10.2. Chromatographic cross-fractionation results of a *block*-copoly(styrene/methyl methacrylate) sample by preparative adsorption chromatography on silica and subsequent SEC analysis of the fractions. The *arrows* indicate apex positions. From Ref. [8] with permission

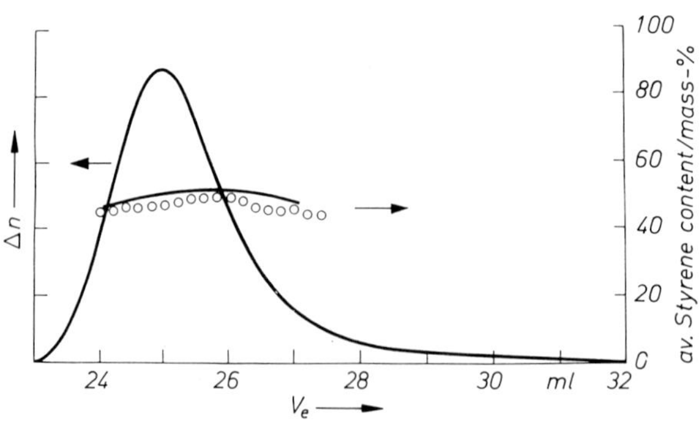

Fig. 10.3. SEC curve with point-by-point evaluation of the styrene content (same sample as in Fig. 10.2). From Ref. [9] with permission

The CCD curves of complex copolymers can be reliably derived by cross fractionation and totalling the data over all chain lengths. In fact, this is a reduction of the information available but it makes sense if comparison of CCF results with calculated CCD curves is aimed at.

10.3 Chromatographic Cross-Fractionation of Model Mixtures

Investigations of this kind demonstrate the efficiency of gradient HPLC for separation by composition and are useful for the quantification of MW effects. With suitable copolymer standards, CCF of model mixtures can provide the calibrations required for the analysis of unknown distributions in real copolymers. Examples of CCF of model mixtures have been presented already in Chap. 6 (see, e.g. Figs. 6.7, 6.8, and 6.14).

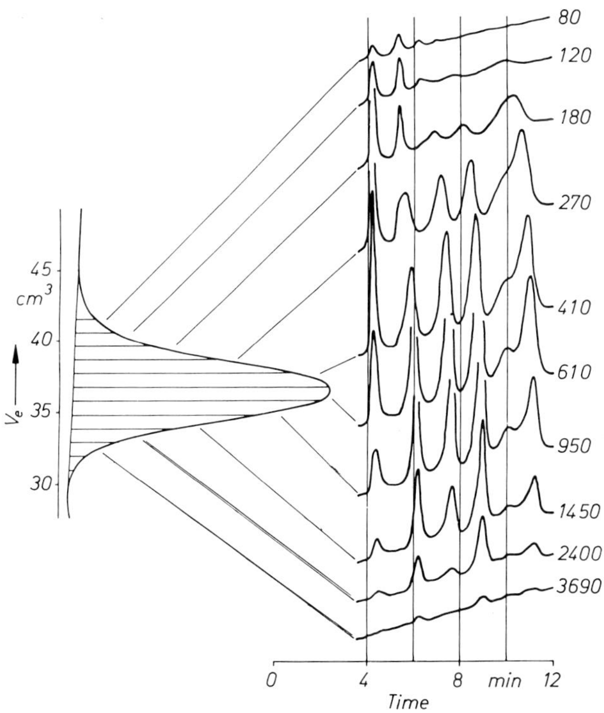

Fig. 10.4. Chromatographic cross-fractionation of the mixture of five *stat*-copoly(styrene/acrylonitrile) standards by SEC and gradient HPLC on CN bonded-phase through *iso*-octane/THF (50–100% in 10 min). SEC conditions: five columns (330 × 7.8 mm each) packed with μStyragel, $d_0 = 50$ nm, 100 nm, 1000 nm, 10^4 nm, or 10^5 nm; eluent: THF stabilized with 0.025% *tert*-butylcresol, flow rate 1 ml/min; $V_0 = 0.5$ ml, $m_0 = 1.04$ mg. HPLC conditions: column 150 × 4.6 mm; $d_0 = 6$ nm; $d_P = 5$ μm, 50 °C, flow rate 1 ml/min; UV detection at 259 nm; MW of the fractions indicated as 10^{-3} *M*; copolymers: A-E (Table 9.1), content in the starting mixture (in the order of elution): 11.5 mass% A, 22.5% B, 11.7% C, 26.1% D and 22.2% E. By courtesy of Elsevier Applied Science Publishers Ltd. Barking, UK [11]

10.3.1 Model Mixtures of *stat*-Copoly(Styrene/Acrylonitrile)

The first published separation of this kind is shown in Fig. 1.2. In order to precisely evaluate both MW and CC calibration [11] for subsequent analyses of commercial S/AN products, CCF was accomplished on a mixture of five model copolymers (containing 16, 23, 29, 36, or 43% AN, in the order of HPLC elution), see Fig. 10.4. SEC was performed on a bank of five μStyragel columns with eluent THF. The size of the starting sample was 1.04 mg, dissolved in 500 μl THF.

About 1 ml SEC eluate each yielded a fraction. The eluate portions were collected in vials fitting to an autoinjector. In the following gradient HPLC, aliquotes of 200 μl of each fraction were injected without any pretreatment. The sample size in gradient HPLC was estimated from the relative area of the respective SEC slice and the total amount of starting material.

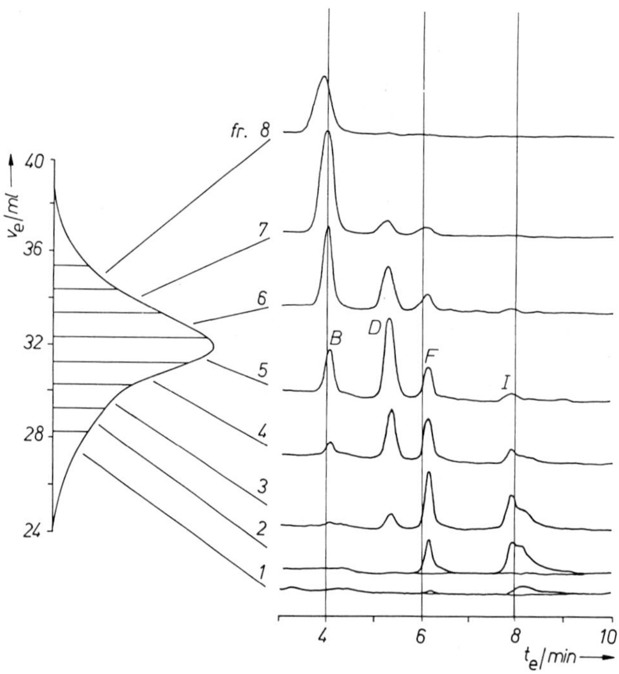

Fig. 10.5. Chromatographic cross-fractionation of the mixture of four *stat*-copoly(styrene/2-methoxyethyl methacrylate) standards by SEC and gradient HPLC. SEC conditions: two mixed-gel columns GMH6 (Toyo Soda, Japan), each 600 × 7.8 mm, $d_P = 8$–$10\,\mu m$, eluent: THF analytical grade, flow rate 1 ml/min, $V_0 = 0.2$ ml, $m_0 = 1.118$ mg. MW of the SEC fractions ($10^{-3}M$): *1*: 408, *2*: 217, *3*: 144, *4*: 93.8, *5*: 60.3, *6*: 38.4, *7*: 24.5, *8*: 15.5. HPLC conditions: CN bonded-phase column (60 × 4 mm; $d_0 \leq 5$ nm, $d_P = 5\,\mu m$), 50 °C, gradient *iso*-octane/methanol (2–52% in 10 min) with 30% THF throughout ("sudden-transition gradient"), flow rate 0.5 ml/min. Copolymers (in the order of elution): B, D, F, and I, see Table 9.8. Chromatograms monitored by UV detection at 230 nm. By courtesy of John Wiley & Sons, Ltd. [12]

10.3.2 Model Mixtures of *stat*-Copoly(Styrene/2-Methoxyethyl Methacrylate) Samples

For SEC, a set of two PS-gel columns was used. Figure 10.5 shows the SEC curve monitored by a refractive index detector after the injection of 1.118 mg of the mixture of four S/MEMA copolymers (containing 26, 49, 62, or 87% MEMA, in the order of HPLC elution). Eluate fraction of 1 ml were collected under a cover of helium in 2.5 ml vials suited for the autosampler of the HPLC apparatus. The vials were capped immediately after filling. The right-hand part of Fig. 10.5 presents the UV tracings of subsequent HPLC analyses of the fractions.

From Table 9.8 it can be read that, with the specimens B, D, F, and I, molecular weight increased with MEMA content. Interference from KMH data excluded, SEC fractionation of a mixture of that kind should yield separation by molecular size which, at the same time, is a prefractionation by composition.

Figure 10.5 shows that the experimental result [12] fulfilled this expectation. The recordings from gradient HPLC conceil a high content of sample B in the low MW fractions no. 8, 7, and 6, of samples D and F in the middle of the MW distribution, and of sample I in the high MW fractions no. 3, 2, and 1.

10.4 Chromatographic Cross-Fractionation of Real Copolymers

10.4.1 Investigation of a Commercial *stat*-Copoly(Styrene/Acrylonitrile) Sample

10.4.1.1 Calibration. For the calibration of a HPLC system with respect to composition and MW, the first moment (i.e. the weight average) of the peaks in Fig. 10.4 were estimated through point-by-point reading the difference between the HPLC curve and the experimental baseline of a blank gradient [11]. From these t_e-data, the gradient program, and the gradient dwell time, the volume fraction of THF, Φ, could be evaluated which eluted a copolymer of composition w. This procedure was applied to all SEC fractions yielding a set of $\Phi(M, w)$ data. The MW effect was quantified by plotting Φ vs $M^{-0.5}$ as described in Sect. 8.5.1.

The slope factor C_2 in Eq. (5.3) was found to be slightly dependent on acrylonitrile weight fraction w_{AN} ($= w$) of the copolymer: $C_2 = 1716 + 2040 w_{AN}$. For the sample to be investigated, w_{AN} was within the limits 0.31 and 0.23, see Fig. 10.6. Hence, C_2 fell in the range $C_2 = 2350$ to $C_2 = 2190$ (for the upper and lower composition limit, respectively) with $C_2 = 2255$ for the main portion of the sample (26.4% AN). In a first approximation, the latter value of C_2 was used in the MW correction of the whole sample.

All data were referred to $M = 140{,}000$ which is near to the number average of the sample. The volume fraction $\Phi(140, w)$ was uses as a reference value. It was calculated from the individual $\Phi(M, w)$ data measured with SEC fractions of molecular weight M by the relation

$$100\,\Phi(140, w) = 100\,\Phi(M, w) - 2255\left(\frac{1}{\sqrt{140.10^3}} - \frac{1}{\sqrt{M}}\right) \qquad (10.1)$$

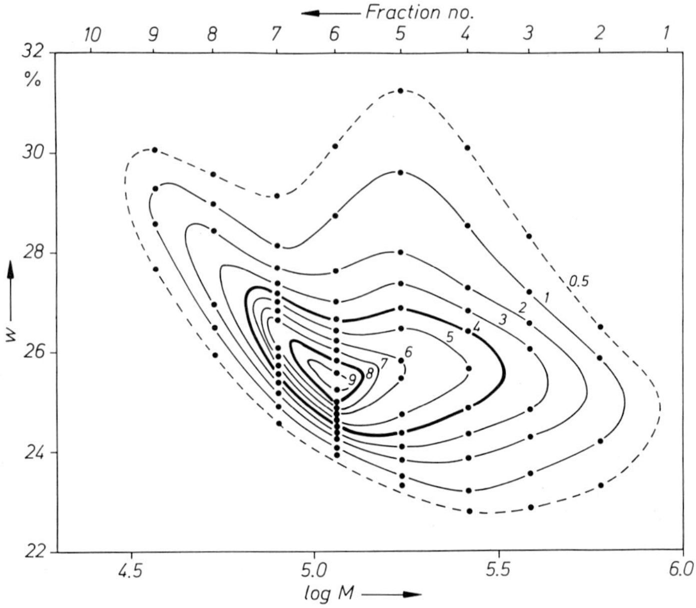

Fig. 10.6. Contour-line map of the distribution in MW and AN content of a commercial S/AN copolymer according to chromatographic cross-fractionation by SEC and gradient HPLC. SEC and HPLC conditions as in Fig. 10.4, but $m_0 = 0.625$ mg. Each row of dots is from HPLC measurement on a respective SEC fraction whose number is indicated at the scale on top of the figure. The contour lines connect points of equal height in the HPLC curves. By courtesy of Elsevier Applied Science Publishers, Ltd., Barking, UK, [11]

(Note that $\Phi(M, w)$ indicates the volume fraction of THF in mixtures with *iso*-octane. Equation (10.1) is the quantification of the fact that, at a given copolymer composition, more THF is required for the elution of a high-MW sample than for a low-MW one.)

In the S/AN-eluent system investigated, the sample composition effect read:

$$100\,\Phi(140, w) = 39.015 + 117.8 w_{AN} \qquad (10.2)$$

i.e. the higher the AN content was, the more THF was required for eluting that sample portion. With the help of this relation, the composition w_{AN} at any point of the HPLC curve could be evaluated through

$$w_{AN} = 0.849\,\Phi(140, w) - 0.331 \qquad (10.3)$$

(correlation: $r^2 = 0.9869$).

10.4.1.2 Investigation of a Commercial Sample. All chromatograms required either for calibration or in the cross fractionation of a commercial sample were measured in an uninterupted series of analyses with the HPLC apparatus running round the clock.

The commercial sample was repeatedly fractionated by SEC using 0.5 ml of a 0.125 mass% sample solution. Two sets of SEC fractions were investigated through gradient HPLC, one precisely in the manner described with reference to Fig. 10.4, the other one (which, unfortunately, was not complete) in an almost identical way but on a RP C18 column. A third set of SEC fractions was analysed by pyrolysis gas chromatography.

Figure 10.6 presents the contour-line map drawn on basis of HPLC measurements of SEC fractions from 0.625 mg starting material. A CN bonded-phase column was used. The chromatograms were eventually normalized to the relative area of the respective SEC slices and evaluated with the help of the calibrations. The points in Fig. 10.6 represent the AN content where the measured distribution curve reached a given level (which is indicated at the contour lines in arbitrary units).

10.4.1.3 Reliability. The average AN content in each SEC fraction was calculated from the height of the distribution curve, h_i, at the respective value of AN content, w_i, by $w_{av} = (\sum h_i \times w_i)/\sum h_i$. This data enabled the chromatographic results to be compared with other measurement on the same sample. A synoptic presentation of the results is given in Fig. 10.7.

The circles show the results of pyrolysis gas chromatography of the fractions. Here, the polymer of SEC fractions was isolated under a cover of nitrogen by

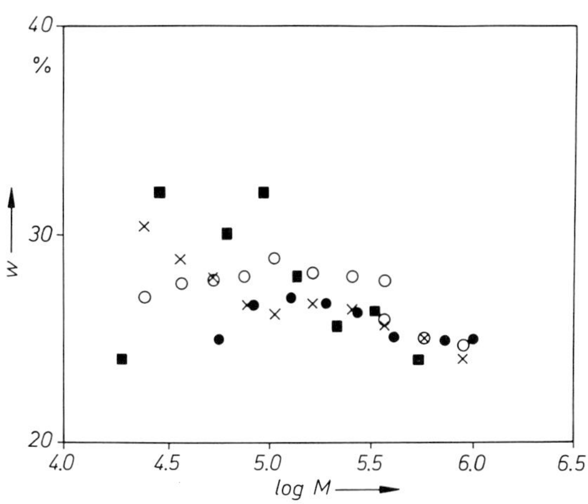

Fig. 10.7. Average acrylonitrile content of SEC fractions as measured by four independent methods. Comparison of CCF results with dual-detection SEC (filled squares) and pyrolysis gas-chromatography (PGC, circles). ×: CCF on CN column, ●: on C18 column. PGC conditions: about 10 μg S/AN in acetone were injected with a split ratio 1:40 into a Perkin Elmer gas chromatograph, type F17, equipped with a 0.25 mm SS capillary column, 85 m in length, coated with Carbowax 1540. Temperature program: 12 min at 40 °C, followed by a rise to 90 °C within 10 min and completed by 20 min at 90 °C. A Fisher high frequency Curie-point pyrolysator type 310 was used at 600 °C for 10 s. Carrier gas: helium. linear velocity 30 cm/s. By courtesy of Elsevier Applied Science Publishers, Ltd., Barking, UK, [11]

vaporizing the THF. The dried residues were redissolved in acetone. About 10 μg polymer of each fraction were injected at a split ratio 1:40 and chromatographed on a capillary column with a temperature program. The AN content was calculated from the ratio of styrene and acrylonitrile peak area by comparison with the corresponding ratio of a reference S/AN copolymer.

Another attempt to evaluating the AN content at different points of the SEC elution curve was made by dual-detection SEC using a differential refractometer and a UV detector at 254 nm. From date given in literature [13, 14] it follows that the UV extinction coefficient ε_s of the styrene portion in a S/AN copolymer can be represented as a function of the weight fraction of styrene, w_s, by

$$\varepsilon_S = \varepsilon_{PS}[1 - 0.28(1 - w_S)/w_S] \tag{10.4}$$

where ε_{PS} denotes the molar absorptivity of PS.

The refractive index increment also is a function of copolymer composition:

$$\frac{dn}{dc} = 0.070 + 0.125 w_S \tag{10.5}$$

With (i) the normalized refractive index tracing, $p = f(V_e)$, (ii) the normalized UV recording, $q = f(V_e)$, (ii) the styrene content $w_{S,o}$ of the whole copolymer, and (iv) Eqs. (10.4) and (10.5), the styrene content w_i of each elution volume could be calculated from the chromatogram values p_i and q_i by

$$w_i = \frac{C_1 + C_2 p_i/q_i}{C_3 + C_4 p_i/q_i} \tag{10.6}$$

where $C_1 = 4.57 w_{S,o} - 1$, $C_2 = 1.79 w_{S,o} + 1$, $C_3 = -8.18 w_{S,o} + 1.79$, and $C_4 = 8.18\, w_{S,o} + 4.57$. Results obtained this way are indicated by squares in Fig. 10.7.

Figure 10.7 shows that the results of four independent investigations agreed sufficiently, at least among fractions of high MW. In total, the results calculated from HPLC measurements scatter less than those from conventional analyses. Thus, CCF technique deserves at least the same credit as methods used in routine work.

10.4.2 Chromatographic Cross-Fractionation of a *stat*-Copoly(Styrene/Methyl Methacrylate) Sample in two Independent Ways

Recently, Mori [15] reported on CCF of a S/MMA copolymer by (i) prefractionation through gradient HPLC and subsequent SEC analysis of the fractions as well as by (ii) SEC prefractionation and investigation of the fractions by gradient HPLC. The sample ("H-3") was polymerized in bulk at 60 °C from a monomer mixture (containing 68.4 mass% MMA) on the addition of 0.15% azobis(isobutyronitrile) as an initiator to a conversion of 97.6%.

In both separating strategies, SEC was performed on a bank of two PS-gel columns in eluent TCM/EtOH (99:1). Gradient HPLC was carried out at 30 °C on a silica column. The samples were dissolved in a portion of the starting eluent and injected into TCM/EtOH (99:1) 1 min after the start of a gradient program. The gradient was essentially the linear increase of EtOH concentration from 1 to 4.5% in 15 min followed by an isocratic period of 10 min at 4.5%.

10.4.2.1 Calibration

1. Gradient HPLC: The effect of MMA content on elution volume was evaluated by using the S/MMA copolymers J, K, L, and M (see Table 9.3) as calibrating standards and measuring their retention under HPLC conditions as described above.
2. SEC: elution in TCM/EtOH (99:1) was calibrated by polystyrene standards of narrow MW distribution. With this calibration, a M_{PS} value could be estimated at any experimental elution volume V_i. The MW value of a copolymer with MMA mole fraction x could be calculated from the PS equivalent MW by

$$M_{COP} = (1 - x)M_{PS} + xM_{PMMA} \tag{10.7}$$

where

$$M_{PMMA} = 1.967 M_{PS}^{0.918} \tag{10.8}$$

(which had been measured in THF [16]).
3. Since the refractive index increment of a copolymer depends on composition, the height of the SEC curve monitored with samples of different MMA weight fraction w was corrected by factor $f = 1 + 1.18w$.

10.4.2.2 CCF of S/MMA through Gradient HPLC Prefractionation and SEC. Figure 10.8 shows the separation into three fractions by gradient HPLC of 120 μg of the high-conversion sample "H-3". Redrawing the elution curve of Fig. 10.8 with the help of an adequate calibration yielded the differential CCD curve, Fig. 10.9. (In the absence of any MW effect in gradient HPLC, the transformation is correct.) Figure 10.9 indicates that the sample consisted of molecules with MMA content between 54 and 85%. For SEC investigation (and

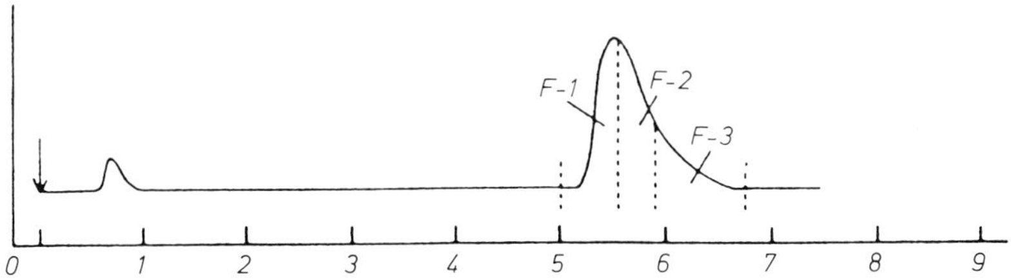

Fig. 10.8. Fractionation of *stat*-copoly (styrene/methyl methacrylate) sample "H-3" (67.2 mass% MMA, conversion 97.6%) by gradient HPLC on a silica column (50 × 4.6 mm; $d_o = 3$ nm; $d_p = 5$ μm) at 30 °C. Eluent **A**: chloroform-ethanol (99:1), eluent **B**: TCM-EtOH (95.5:4.5), gradient: 0–100% **B** in 15 min, then another 10 min at 100% **B**, flow rate 0.5 ml/min, $V_o = 100$ μl, $m_o = 120$ μg. From Ref. [15] with permission

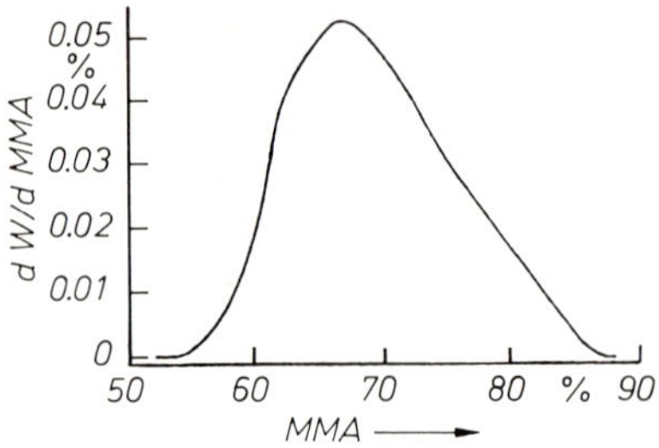

Fig. 10.9. Differential composition distribution curve of S/MMA copolymer "H-3". From Ref. [15] with permission

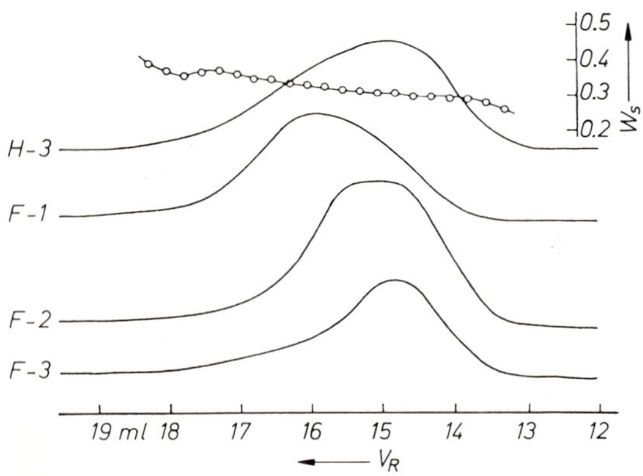

Fig. 10.10. Size-exclusion chromatography of copolymer "H-3" and fractions 1-3 obtained through gradient HPLC (see Fig. 10.8). SEC in chloroform-ethanol (99:1) on a bank of two Shodex polystyrene-gel columns (250 × 8 mm each), flow rate 1 ml/min, $V_o = 0.1$ ml, $m_o = 0.2$ mg [17]. From Ref. [15] with permission

the determination of composition by IR absorption), the HPLC separation of 120 μg each was repeated 50 times and respective fractions were combined.

Analytical SEC was performed on two PS gel columns 250 × 8.0 mm. Figure 10.10 displays the SEC elugrams of the HPLC fractions indicated in Fig. 10.8. With increasing fraction number, the SEC elution curves in Fig. 10.10 shifted towards lower elution volume, i.e. higher MW. An increase in MW can be also read from Table 10.1 which compiles average MW data calculated from the SEC curves. ("The values in the table are all the polystyrene equivalent molecular weights" [15], p. 1126).

The chemical composition of each fraction was measured by the absorbance of phenyl groups at 699 cm^{-1} and of carbonyl groups at 1730 cm^{-1}. The total of the MMA content of the fractions (67.3 mass%) meets almost exactly the directly measured value of the starting material (67.2%). In contrast to this, the MW averages M_n and M_w calculated from the data of the fractions ($10^{-3}M_n = 403$; $10^{-3}M_w = 743$) deviate substantially from the values of the unfractionated sample, 362 and 838, respectively. (Recalculation of the MW-data with the help of Eqs. (10.7) and (10.8) gives almost no improvement.)

Table 10.1. Composition and molecular weight of a high-conversion styrene/methyl methacrylate copolymer ("H3", see Table 9.3) and its fractions obtained by gradient HPLC (data from Ref. [15])

Fraction No.	Mass (weight fraction)	Composition MMA (mass%)	$10^{-3}\,M_n$	$10^{-3}\,M_w$
1	0.321	63.4	287	500
2	0.420	68.3	476	842
3	0.259	70.6	428	882
Total	1.000	67.3	403	743
Raw sample (S/MMA, polymerized from a monomer mixture containing 68.4 mass% MMA: conversion 97.6%):		67.2	362	838

10.4.2.3 CCF of S/MMA by SEC Prefractionation and Subsequent Gradient HPLC. Preparative SEC was performed on a set of two PS-gel columns (500 × 8.0 mm). Injections of 0.25 ml containing 0.5 mg copolymer each were repeated more than ten times and corresponding fractions collected [17]. Gradient HPLC of these fractions yielded tracings similar to that in Fig. 10.8. They were converted into CCD curves by calibration, see Sect. 10.4.2.1, paragraph (1). In Fig. 10.11, the CCD curves of all SEC fractions are drawn at the position of the latter on the axis of SEC elution volume, i.e. with MW increasing from the furthest-forward curve to those behind. With these results and the copolymer MW calculated through Eq. (10.7), the contour-line map of the distribution both in MW and composition was drawn, see Fig. 10.12.

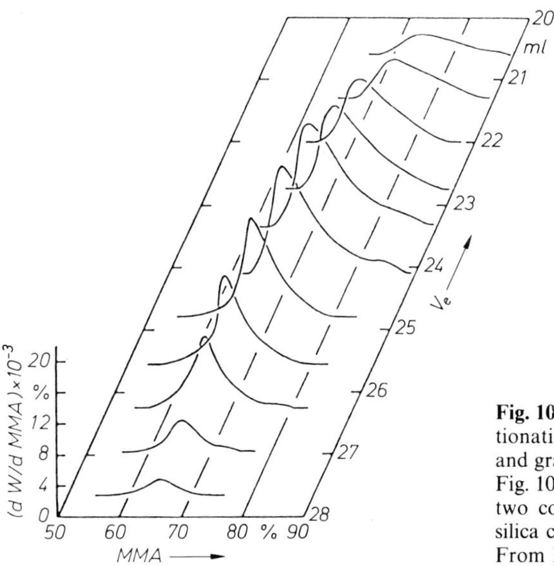

Fig. 10.11. Chromatographic cross-fractionation of S/MMA sample "H-3" by SEC and gradient HPLC. SEC conditions as in Fig. 10.10, but $V_0 = 0.25$ ml, $m_0 = 500\,\mu g$ two columns 500 × 8.0 mm [17]; HPLC: silica column and gradient as in Fig. 10.8. From Ref. [15] with permission

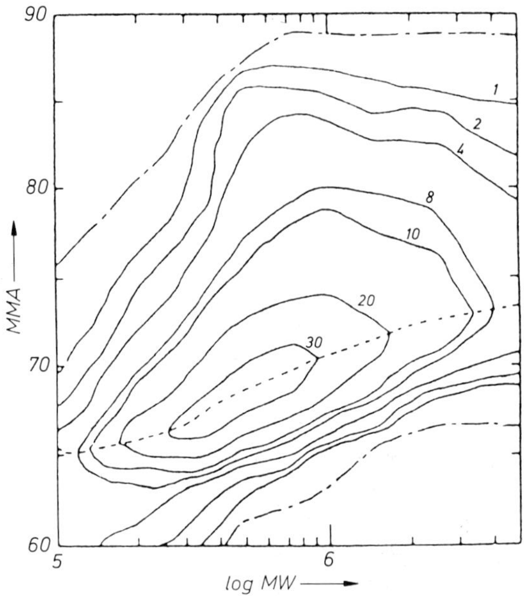

Fig. 10.12. Contour-line map of S/MMA sample "H-3". From Ref. [15] with permission

10.4.3 Investigation of *stat*-Copoly(Styrene/2-Methoxyethyl Methacrylate) Samples of Different Degree of Conversion

The objective of this study was the distribution analysis of two S/MEMA copolymers which were polymerized from the same batch to different degree of conversion. The initial monomer mixture contained less MEMA (23.3 mass%) than an *azeotropic copolymer* (59.0%) and would yield, at infinitesimal conversion, a copolymer containing 33.1 mass% MEMA. One sample (coded "J") was polymerized to 10.1 mass% conversion and contained 32.3% MEMA ($M_w = 37{,}200$), the other one ("K"), polymerized to 86.4% conversion, had 26.0% MEMA and $M_w = 57{,}200$. The polymerization was performed in benzene solution with azobis(isobutyronitrile) (AIBN) as an initiator.

10.4.3.1 Calibration. The information comprised in Fig. 10.5 enables the elution time in gradient HPLC to be evaluated as a function of MEMA content and MW. The dependence of elution time on copolymer composition must be ensured by calibration at least in that range of MW which is representative for the copolymer to be investigated.

The first step was the search for a possible MW effect [12]. In Fig. 10.13, the first moment of the peaks in Fig. 10.5 is plotted vs $M^{-0.5}$. The first moment is expressed in terms of elution time (right scale) or volume fraction *iso*-octane (gradient component **A**), left scale. (The latter mode of plotting has the advantage that slope factors can be compared directly for any gradient rate used).

Least-square evaluation of the straight lines in Fig. 10.13 yields $100\,\Phi_{\text{ioct}} = 58.4 + 210/M^{-0.5}$ for sample B and similar data for the other

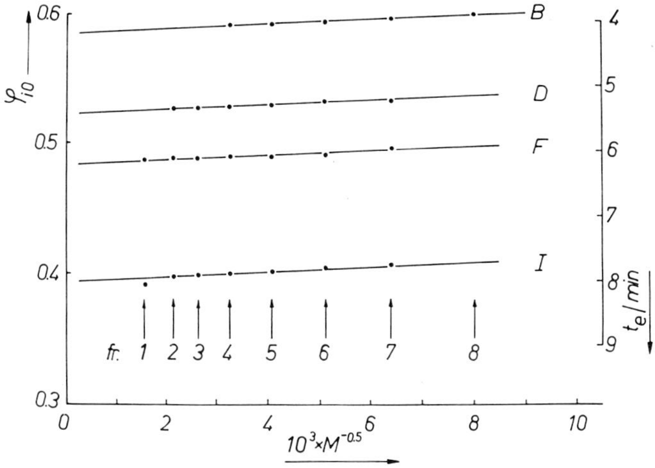

Fig. 10.13. Molecular-weight effect in gradient HPLC of *stat*-copoly (styrene/2-methoxyethyl methacrylate). By courtesy of John Wiley & Sons, Ltd., [12]

specimens, see Table 8.1. The small values of slope factor are remarkable as is the fact that in the system investigated composition had almost no influence on the MW effect.

With this data, the composition effect on retention could be estimated for any MW in the limits of experimental evidence. Figure 10.14 shows a plot of that kind for $M = 40{,}000$. The relation is given as the dependence of elution time (right scale) on MEMA content; the line is determined by four heavy points, due to the samples used in Fig. 10.5. The open circles are read from Fig. 9.21 and refer to volume percentage THF (left scale), which is related to t_e by the gradient program.

The reason for this simultaneous presentation of data measured under different conditions was the question of whether or not the calibration based on four copolymers could be assumed to hold also for further samples. This can be answered positively because both curves have analogous slope. Surprisingly enough, both lines fall almost together when the position of the plot is adjusted by the *i*Oct content in the system, which is either $100(1 - \Phi_{THF})$ (in Fig. 9.21) or $100(1 - \Phi_{THF} - \Phi_{MeOH})$ (in Fig. 10.5). Of course, the proximity is only incidental since, in the separations shown in Fig. 10.5, a different THF level would alter the total of polar eluent components (see Sect. 5.6 and Fig. 5.15). Additionally, the MW value selected for calibration influences the result. Nevertheless, the similarity of the curves in Fig. 10.14 supports the reliability of the calibration applied.

Signal size was evaluated by measuring the height of the elution curve over the experimental baseline at equidistant values of elution time. About 20–60 data points per peak have been taken. The total of these height data of a given peak was used as a measure of its size.

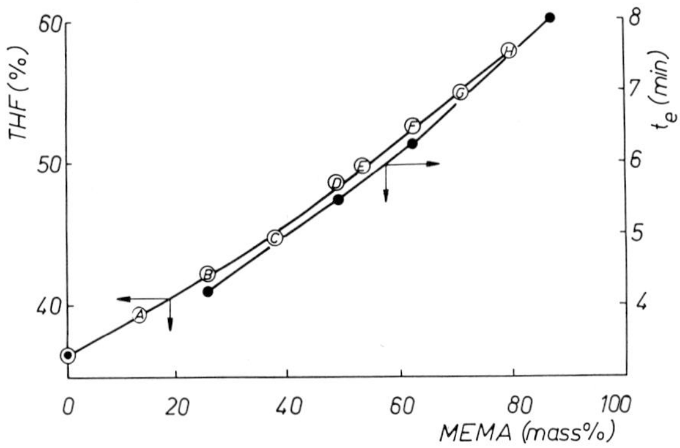

Fig. 10.14. Calibration curve for the gradient HPLC of S/MEMA copolymers (MW 40,000, •—•). (For comparison, the elution characteristics are added which can be derived from Fig. 9.21: circles with indication of sample code, sample data see Table 9.8)

In order to find out the effect of copolymer composition on detector signal under the conditions of CCF, the total of the HPLC peak areas of sample B in all SEC fractions (Fig. 10.5, fr. 1–8) was separately calculated. The same totalling was performed with the peaks of samples D, F, and I. The sums were subdivided by the amount of each copolymer in the starting mixture.

The values of specific peak area ^1A estimated from Fig. 10.5 are plotted vs copolymer composition in Fig. 10.15. The first impression is a slight decrease of specific peak area with increasing MEMA content. The reason for this only small dependence is the fact that the chromatograms were monitored at 230 nm where MEMA units substantially contribute to UV absorption. The higher value of sample I is unexpected. Fortunately, the position of this point has no influence on the investigation of the samples J and K whose composition range is indicated by bars in Fig. 10.15. (Only the position of the bars with respect to the MEMA axis bears relevance; their position relative to the ordinate is arbitrary). For copolymers J and K, the dependence of specific absorptivity on composition can be assumed to follow the dashed line in Fig. 10.15. It amounts to about 4% decrease in detector signal by 10% increase in MEMA content. In a first approximation, this relatively small composition effect was neglected.

Since the signal size of the calibrating standards in Fig. 10.5 is almost independent of copolymer composition, the sum of all peak areas in a given SEC fraction should be a measure of the amount of polymer in this fraction. The latter was estimated independently from the relative area between the SEC curve, the SEC baseline, and straight orthogonal lines which indicate how the fractions had been cut, see the left-hand part of Fig. 10.5. Figure 10.16 shows that, with a sufficient correlation, the relative size of the HPLC peaks (signal at 230 nm) varies linearly with the relative SEC area (RI signal).

If solvent evaporation had occurred in the handling of certain fractions or if differing volumes had been injected into the HPLC apparatus the area of the

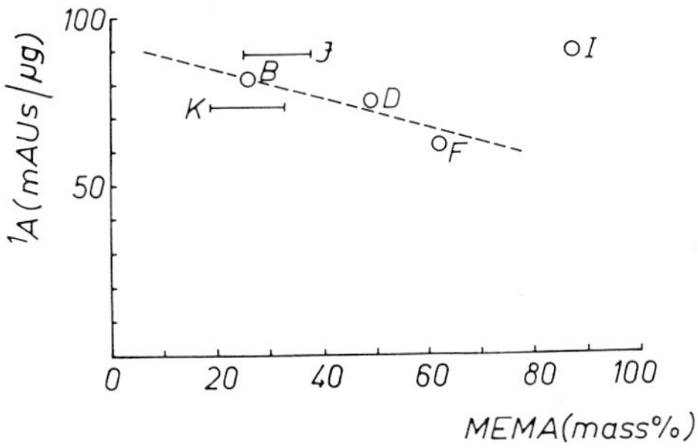

Fig. 10.15. Specific peak area at 230 nm vs composition of the respective S/MEMA calibrating sample. Data from Fig. 10.5. By courtesy of John Wiley & Sons, Ltd., [12]

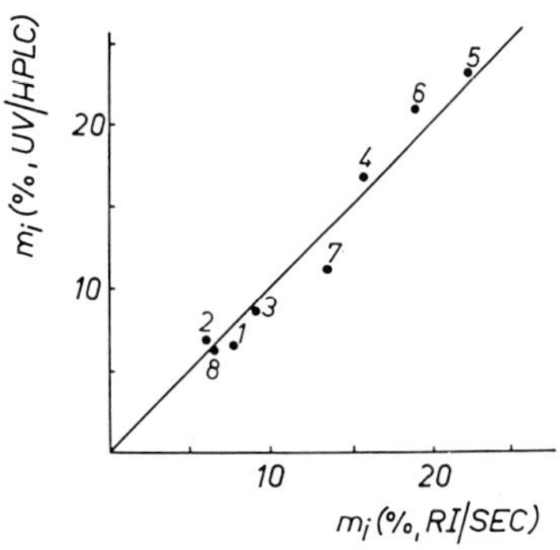

Fig. 10.16. Amount of S/MEMA copolymer detected in gradient HPLC vs sample size: plot of the relative area of UV peaks from HPLC tracings of fractions *1–8* vs relative area of the respective slices under the SEC curve (RI signal). Data from Fig. 10.5. By courtesy of John Wiley & Sons, Ltd., [12]

HPLC peaks would need correction. Figure 10.16 proves that empirical corrections were not necessary; it also once more supports the fact that, in the range investigated, the UV signal at 230 nm is almost independent of SEMA content.

10.4.3.2 Chromatographic Cross-Fractionation of S/MEMA Copolymers J and K. The initial step was SEC, which was performed on 1.15 mg sample material under the conditions described in Sect. 10.3.2.

In order to ensure the closest correspondence between calibration and measurements, the SEC fractions J1–J8 and K1–K8 were, with the help of an autoinjector, investigated together with the fractions fr.1–fr.9 from a calibrating mixture (see Fig. 10.5) and several blank gradient runs in an uninterrupted series of injections under constant elution conditions within 8 hours (for details, see Ref. [12]).

The first group of injections was from fractions in the centre of the SEC elution curves, where most of the sample material was concentrated. These injections were performed with 50 μl volume. The rest was from the slopes of the SEC curves and contained less polymer. Here, injection volume in gradient HPLC was set to 100 μl. All aliquotes were injected without any pretreatment of the collected SEC eluate.

The HPLC tracings of SEC fractions from specimens J and K were converted into composition distribution curves with the help of appropriate calibrations. At stepwise increased ordinate values, the composition was read from these curves and plotted vs the logarithm of MW of the respective fraction. Each column of points in Figs. 10.17 and 10.18 is derived from a given SEC fraction. Contour lines were obtained by connecting experimental points read at a certain height from all CCD curves. The figures at the contour lines are in arbitrary units and reflect the ordinate values which are determined by (i) the amount of material with a given MEMA content and MW, by (ii) the attenuation of the detector signal, and (iii) the format of the HPLC plots [18].

Comparing the contour maps of the low-conversion sample J (Fig. 10.17) and of the high-conversion sample K (Fig. 10.18) one realizes the following features:

The low-conversion sample J has a broader CCD in its low-MW region and a narrower one in its high-MW area. This can be understood as a chain-length effect on CCD. According to theory, the width of the instantaneous CCD decreases with increasing degree of polymerization, see Eq. (2.12).

The high-conversion sample K is richer in styrene than the low-conversion one, i.e. the contour map of the former extends into the area below the map of sample J. This is a consequence of the preferred consumption of MEMA monomers and the extended duration of the polymerization reaction: as the copolymerization proceeds, the monomer mixture becomes poor in MEMA.

The extension towards higher styrene content is predominant in the high MW-region of sample K. This indicates that high-MW species were formed in the

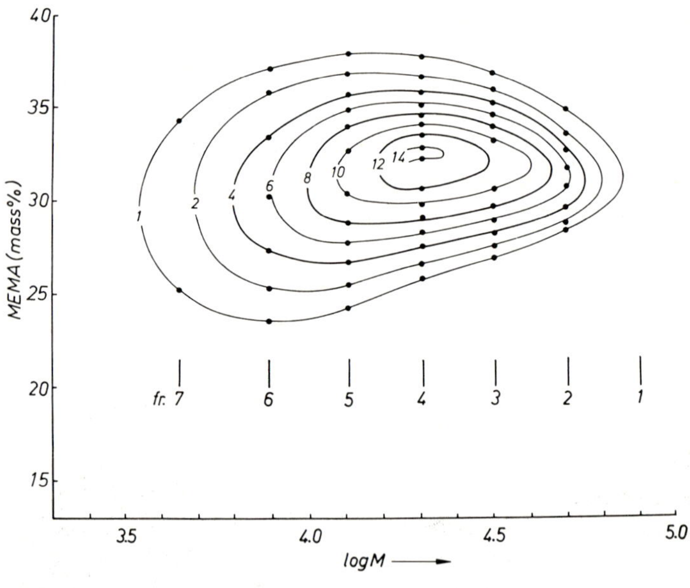

Fig. 10.17. Contour-line map of low-conversion S/MEMA copolymer "J". Points read from composition profiles, which were obtained by gradient HPLC of SEC fractions. The figures at the contour lines indicate the height of the respective reading. By courtesy of Hüthig & Wepf Verlag, [18]

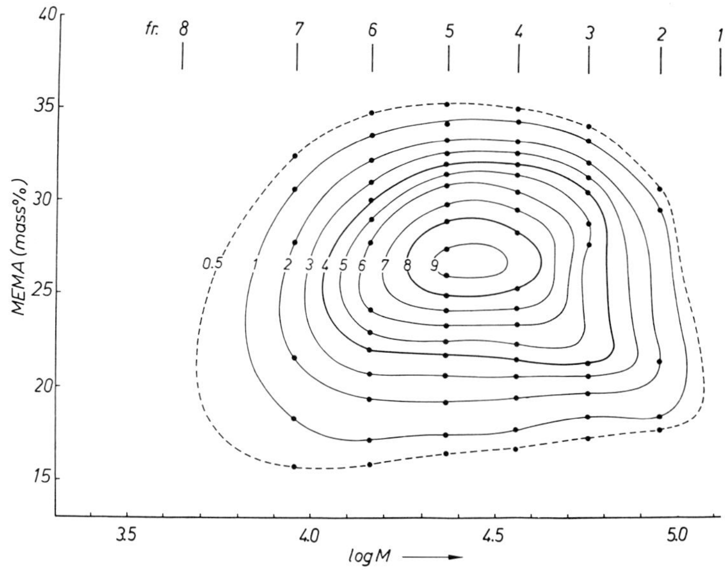

Fig. 10.18. Contour-line map of the high-conversion S/MEMA copolymer "K" (see Fig. 10.17). By courtesy of Hüthig & Wepf Verlag, [18]

final period of the copolymerization. The consequence is a CCD which, in the high-MW region, is even broader than in the lower region. This follows from the contour lines 2 to 9 in Fig. 10.18 representing the main portion of sample K. (Note that this sample no longer shows the apex at about 33% MEMA and log $M = 4.3$ which can be seen in Fig. 10.17. The apex in Fig. 10.18 is definitely lower and located at about 27% MEMA and log $M = 4.4$–4.5).

The experimental background for the extension of the contour lines towards higher MW and lower MEMA content in Fig. 10.18 are the broad HPLC traces found with SEC fractions K2 and K3, see Fig. 10.19. As mentioned above, these

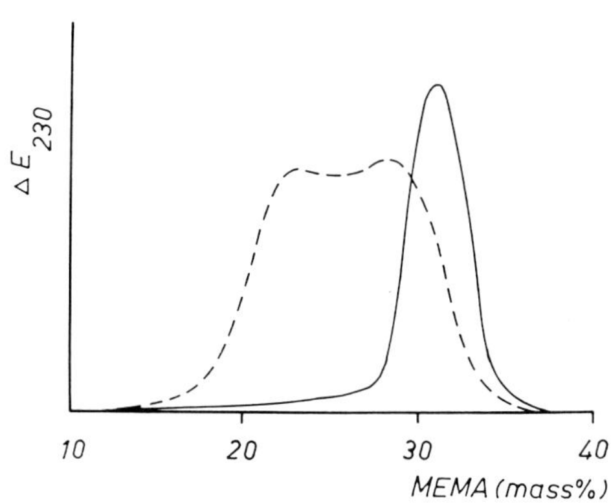

Fig. 10.19. Elution curves (measured by gradient HPLC under identical conditions) of two SEC fractions with almost the same MW. *Full line*: fraction J2 from low-conversion S/MEMA sample "J", MW 50,100; *dashed line*: fraction K3 from high-conversion sample "K", MW 56,600. By courtesy of Hüthig & Wepf Verlag, [18]

fractions had been investigated in a mixed sequence of fractions from either sample J, K, or the calibrating mixture. Only the fractions K2 and K3 showed bimodal peaks. These "suspicious" results appeared again in repeated injections performed the next day.

If the broadening of the CCD were only due to conversion, i.e. due to preferential consumption of MEMA monomers, the extension should be the same in the high-MW and in the low-MW region; i.e. it should parallel the composition axis of the contour map. The diagonal extension in Figure 10.18 reflects an additional effect.

10.4.3.3 Reliability of S/MEMA Cross Fractionation. Integral CCD curves were derived from the experimental results shown in Figs. 10.17 and 10.18 by totalling the amounts of polymer having MEMA content less than (or equal to) the value indicated by the abszissa of the respective point in Fig. 10.20. After each summation, the upper boundary value (corresponding to the abszissa value in Fig. 10.20) was incremented by 1% MEMA and the calculation repeated. The values indicated by *W* give the average composition of the whole sample. These date were calculated from the relative amount of each SEC fraction and the average composition of the latter due to the respective HPLC result.

Figure 10.21 presents the theoretical CCD calculated by numerical integration of Eq. (2.14) over conversion. The *monomer reactivity ratios* were taken as $r_S = 0.46$ and $r_{MEMA} = 0.48$, the MW values as 37,200 (sample J) or 57,200 (sample K). The average composition was measured by the refractive index of the whole copolymer and is indicated at the curves (symbol: *W*).

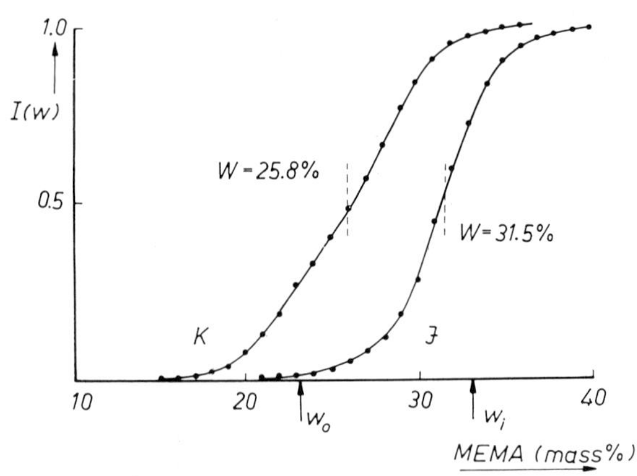

Fig. 10.20. Integral mass-distribution $I(w)$ of chemical composition w (given in mass% MEMA) of samples J and K. Data obtained through step-by-step summation of the results from chromatographic cross-fractionation. The *arrow* with the notation w_0 indicates the composition of the starting mixture, w_i refers to a copolymer obtained from this mixture at infinitesimal conversion. By courtesy of Hüthig & Wepf Verlag [18]

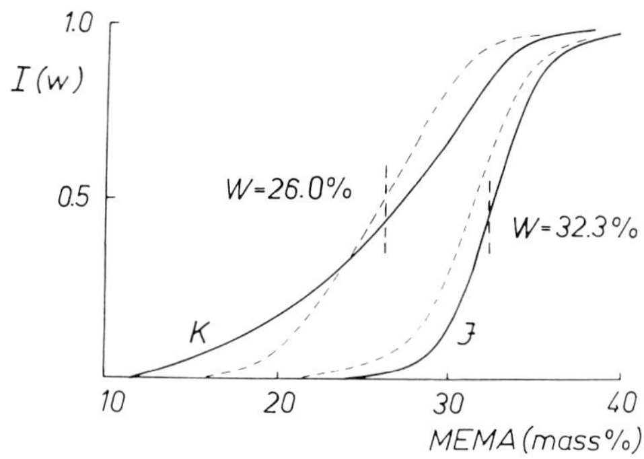

Fig. 10.21. Integral mass-distribution I(w) of chemical composition w calculated from theory for 10.1 and 86.4% conversion and a starting mixture containing 23.3% MEMA. The *dashed curves* repeat the experimental results from Fig. 10.20. By courtesy of Hüthig & Wepf Verlag [18]

Comparing the calculated and experimental values of average composition, one finds slightly lower values for the experimental ones (0.8% for the 10.1%-conversion sample and 0.2% for the high conversion one).

As mentioned in the discussion of Fig. 10.15, the composition effect on detector response was ignored in evaluating the HPLC traces of the SEC fractions. This might have caused overestimation of portions having higher absorptivity values (i.e. of portions richer in styrene) and underestimation of portions having lower absorptivity (i.e. being richer in MEMA). An imperfection like that would cause the average MEMA content to be too low – just as observed. But on the whole, the differences are so small that they also can be looked at as experimental errors.

For the low-conversion sample J, the shape of the experimental curve is in fair agreement with the calculated one, whereas the experimental curve of the high-conversion specimen K differs significantly from the calculated distribution. The discrepancy could be due to insufficient separating power of the chromatographic procedure or to deviation of the copolymerization from the reaction mechanism assumed in calculating the distribution. (The detector sensitivity for samples of differing composition cannot be responsible for the observed deviation because the overestimation of styrene-rich portions and underestimation of MEMA-rich portions would make the experimental curve less steep. This is just the opposite of the observed effect.)

From Fig. 10.21 it can be concluded that the resolution of the experimental set-up was sufficient to measure accurately the CCD of a low-conversion sample. Here, the agreement in shape between experimental and calculated curve could hardly be better. In general, chromatographic separations are more difficult with narrow distributions than with broad ones. A method that works well on a difficult task will scarcely be the reason for unexpected results with less complicated samples.

The equations used for the calculation of CCDs have been derived under the assumption that all chains, regardless of degree of conversion at which they were grown, yield individual macromolecules without chain transfer to dead polymer. For S/MEMA copolymers, this would mean that macromolecules rich in styrene, which are formed at later stages of the reaction, should principally be separable from those formed earlier, which are richer in MEMA. In the present system, almost nothing is known about chain transfer to polymer but, from basic knowledge on polymerization to high conversion and from the course of styrene polymerization, this effect cannot be ruled out with certainty. If it occurred, it definitely would decrease the width of CCD and increase the MW of the sample. The measured difference in MW between sample J and K (37,200 and 57,200, respectively) is beyond experimental error. It could have been caused by either initiator depletion or chain transfer, or both. The half-life time of AIBN is 22 h at 60 °C. After about two half-lives, the initiator concentration can be assumed to be almost zero. Nevertheless, conversion proceeded from 76% at 198 h through 77.1% (285 h) and 79.3% (440 h) to 86.4% (sample K, 820 h). It may be noted that other samples from this experiment had MW values around 37,000 almost independent of conversion in the range 6.6–66.4% (reached after 90 h). Only at higher degree of conversion increased MW of the polymers up to the value given for sample K.

Thus, the assumptions seems to be justified that the deviation in the curves of the high-conversion sample are not in the first instance caused by insufficient efficiency of the CCF technique.

10.5 Cross Fractionation by Temperature-Rising Elution Fractionation and SEC Analysis of the Fractions

This technique is suitable, e.g. for copolymers of ethylene and α-olefins. It has been applied to linear low-density polyethylene (LLDPE) [19–23], to high-pressure LDPE [19–22, 24], and to block copolymers of ethylene and propylene [24].

Temperature-rising elution fractionation (TREF) has been dealt with in Sect. 9.14 as an analytical method. For cross-fractionation, the apparatus have been scaled up to a separation capability of several grams.

Wild and Ryle fractionated samples of 4 g each by preparative TREF and measured short-chain branching (SCB) and MWD of the fractions by IR and SEC, respectively. Data of both SEC and TREF were used for calculating the distribution in MW and SCB.

Figure 10.22 displays results obtained this way for a common LDPE and for a LLDPE sample. Whereas the first one has a unimodal distribution in methyl endgroups and MW, the latter consists of an almost linear portion of very high MW and a portion of lower MW with a broad distribution of short chain branching.

The existence of two portions of different degree of branching showed up also in the elution curves in analytical TREF [20, 22, 23]. Cross-fractionation provided additional information that the both constituents of LLDPE have a

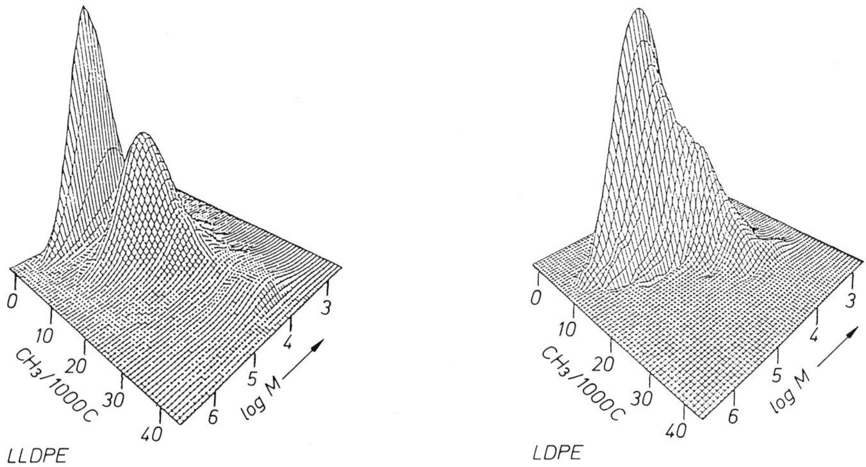

Fig. 10.22. Short-chain branching distribution and molecular-weight distribution of linear low-density polyethylene (LLDPE) and high-pressure LDPE, evaluated by preparative temperature rising elution fractionation (TREF) and SEC analysis of the fractions. TREF conditions: column 508 × 127 mm I.D., packed with diatomaceous earth (Chromosorb-P), sample: 4 g, in hot xylene, cooling rate 1.5 K/h; elution: step gradient, 4 K each. SEC conditions: two DuPont bimodal columns (10 and 100 nm) plus a 400 nm column, eluent 1, 2, 4-trichloro benzene, 145 °C, flow rate 0.7 ml/min, elution monitored by IR detection. From Ref. [21] with permission

pronounced difference in MW. This was confirmed by Usami et al. [23] who performed vinyl end-group analyses of the fractions. The result is shown in Fig. 10.23 together with the analytical TREF curve of the sample. The sudden drop in vinyl-group concentration is due to a corresponding increase in MW.

The Japanese authors also studied the sequence-length distribution in the copolymer by ^{13}C-NMR measurements and calculated the product of *monomer reactivity ratios*, $r_A r_B$. They found for the portion with a broad SCB distribution

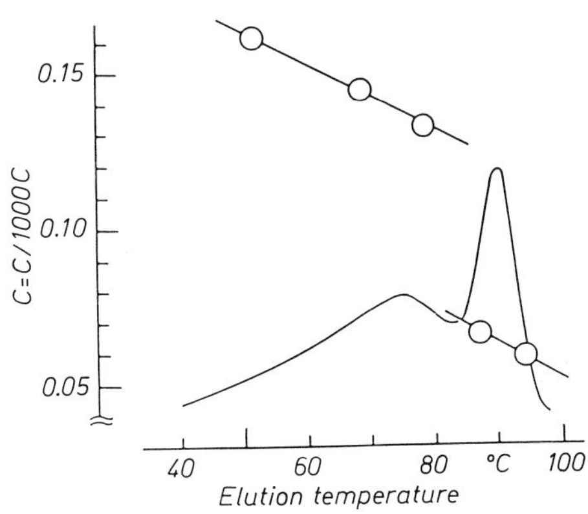

Fig. 10.23. Analytical TREF curve and concentration of vinyl end-groups in fractions from preparative TREF of a LLDPE sample. Vinyl end-groups determined by IR absorption at 910 cm^{-1}. From Ref. [23] with permission

$r_A r_B \approx 0.5$, which is characteristic of an alternating tendency. In contrast to this, $r_A r_B \approx 1$ was found with the high MW portion. This value of the product $r_A r_B$ is characteristic of statistic monomer addition. Thus, the authors suggested the existence of two different active sites in a titanium-based Ziegler catalyst capable of producing LLDPE [23].

The Japanese team also designed an automated apparatus which performed all steps of a cross-fractionation by TREF and SEC in unattended operation

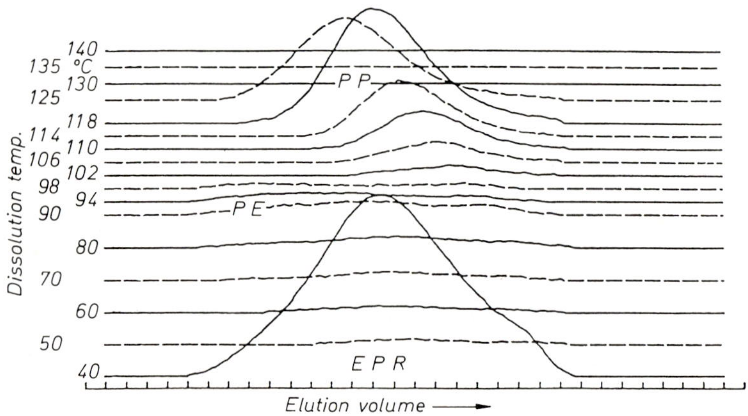

Fig. 10.24. Cross-fractionation of a *block*-copoly (ethylene/propylene) sample by size exclusion chromatography of fractions obtained by TREF (fractionating temperature indicated). From Ref. [24] with permission

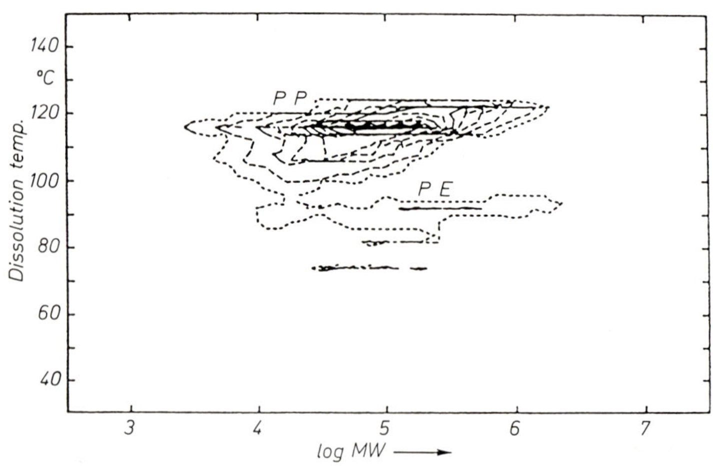

Fig. 10.25. Contour-line map of the homopolymer admixtures in a *block*-copoly (ethylene/propylene) sample. Data from Fig. 10.24. Polypropylene and high-MW polyethylene (both by-products of the desired reaction) can be perceived. From Ref. [24] with permission

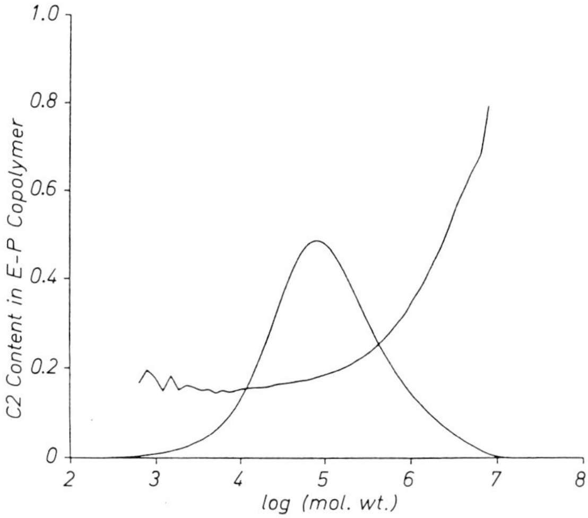

Fig. 10.26. Size exclusion chromatography with IR dual-detection of the sample investigated in Fig. 10.24 and 10.25: apparent MW distribution curve and evaluation of ethylene concentration. From Ref. [24] with permission

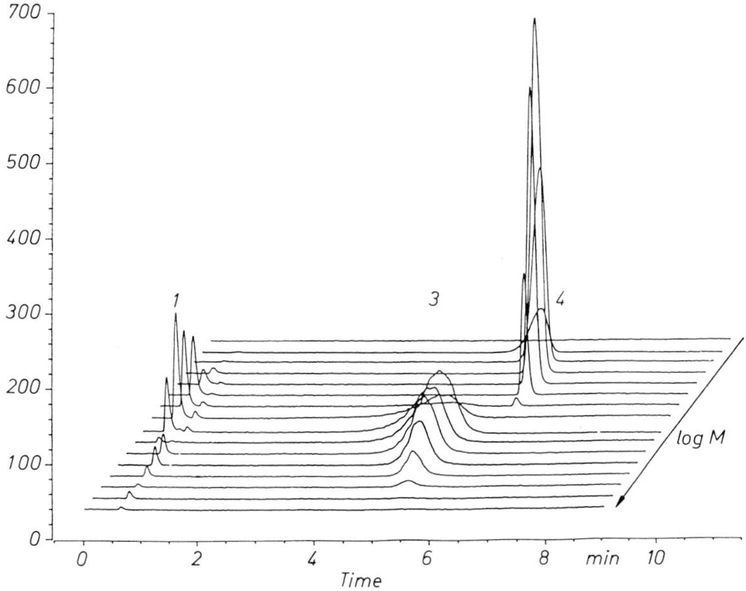

Fig. 10.27. Graft copolymerization of methyl methacrylate onto copoly (ethylene/propylene/diene monomer) (EPDM): Chromatographic cross-fractionation of the reaction product. Gradient HPLC on a CN column (60 × 4 mm; $d_0 \leqq 5$ nm, $d_p = 5$ μm) of SEC fractions obtained from 2 mg starting material. Gradient: *iso*-octane/tetrahydrofuran (1–100% in 6 min), 50 °C, chromatograms monitored by an evaporative light-scattering detector. SEC prefractionation on a bank of two mixed-gel columns (600 × 7.8 mm each) through eluent THF, flow rate 1 ml/min, $V_0 = 0.2$ ml. Peak identification: (*1*) EPDM precursor, (*3*) EPDM-*graft*-MMA, (*4*) PMMA by-product. From Ref. [26] with permission

[25]. Figure 10.24 shows the SEC curves of a sequence of TREF fractions from a *block*-copoly (ethylene/propylene) sample obtained with the help of this "cross-fractionating chromatograph". The final result of this investigation was a computer-drawn contour-line map (Fig. 10.25) which shows the existence of polypropylene and polyethylene, which are by-products of the desired block-copolymerization. It can be clearly seen that the PE has a higher MW. This fully corresponds to an earlier result measured on this sample by dual-detection SEC, see Fig. 20.26: the increase of ethylene concentration with MW can be seen, but SEC analysis could not tell whether the shift was caused by a drift in copolymer composition or by admixed homopolyethylene. Only cross fractionation proved the latter version to be true.

10.6 Chromatographic Cross-Fractionation of Graft Copolymers of Methyl Methacrylate onto Copoly(Ethylene/Propylene/Diene Monomer) (EPDM Rubber)

Gradient HPLC of this copolymer system is dealt with in Sect. 9.12. The same gradient technique has been employed in cross-fractionating a reaction product [26]. The first step was prefractionation by SEC on a bank of two PS-gel columns through THF eluent. 18 fractions were collected. The second step was gradient elution by *i*Oct/THF on a CN bonded-phase column monitored by an evaporative light-scattering detector. Under the chromatographic conditions employed, unreacted EPDM precursor was inertly eluted whereas the desired graft product was retained due to its content in MMA units.

The superposition of the HPLC tracings in Fig. 10.27 gives a perfect impression of the three-dimensional distribution in MW and composition. The reaction product contained the graft product which was eluted at about 5 min. It was found in the SEC fractions no. 2–9. (The numbering runs opposite to the $\log M$-direction. The tracing in the foreground is from fraction no. 1.) Some ungrafted EPDM precursor was eluted at about 0.5 min, which is the elution time of excluded polymer (non-retained sample portions restricted to the interstitial volume of the column). The precursor had a broad MWD and was found in the SEC fractions no. 1–15. The amount was highest in fraction no. 9.

The graft product had its highest concentration in fractions no. 7 and 8. As expected, it had a higher MW than the precursor. With decreasing MW, the graft copolymer was eluted earlier in substantially broader traces. This is presumably not only the reflexion of an MW effect in non-exclusion HPLC. Since retention was mainly due to MMA content, the low-MW fractions must show broader elution curves of the graft copolymer than the high-MW ones.

In the SEC fractions of lower MW, PMMA by-product can be seen which eluted at about 6 min. Only chromatographic cross-fractionation revealed the fact that this by-product is fully isolated from the main product. Neither SEC nor HPLC alone could provide this information, see Figs. 9.28 and 9.29. In both one-dimensional separations, the respective peaks overlapped.

10.7 Chromatographic Cross-Fractionation of Styrene/Vinyl Acetate Block Copolymers

The block copolymers investigated were prepared by free-radical polymerization using polymeric peroxides. Vinyl acetate was polymerized first, then styrene was added and the reaction continued. Samples containing 90, 48, or 49 wt% styrene were obtained from experiments with 90, 70, or 50 wt-% styrene in the feed, respectively. The samples were fractionated [27] by gradient HPLC on a silica column (50 × 4.6 mm, $d_0 = 3$nm, $d_P = 5\,\mu$m) at 30 °C through a gradient DCM/EtOH (2 – 15% in 20 min) into 9 or 10 fractions. The sample load was $m_0 = 800\,\mu$g in $V_0 = 200\,\mu$l. The composition of the fractions was measured by IR absorption, their MW by SEC.

Chromatographic cross fractionation was performed [27] with SEC in the first stage (two columns 500 × 8 mm, packed with polystyrene gel), eluent DCM + EtOH (95:5), $m_0 = 800\,\mu$g in $V_0 = 200\,\mu$l, and gradient HPLC in the second stage. For the latter analyses, the fractions were dried and redissolved in DCM. Gradient HPLC was carried out as described above but with $m_0 = 100\,\mu$g in $V_0 = 100\,\mu$l and UV detection at 254 nm. The results revealed the broad chemical composition distribution of the samples, with VAC content increasing with molecular weight of the fractions.

10.8 References

1. Morgan LW, Jensen DP, Weiss CS (1987) Proceedings ACS Div Polym Materials and Engineering 57: 689
2. Glöckner G, Wolf D unpublished results, S/AN copolymers
3. Teramachi S, Hasegawa A, Yoshida S (1983) Macromolecules 16: 542
4. Belenkii BG, Gankina ES (1977) J Chromatogr 141: 13
5. Belenkii BG, Gankina ES, Nefedov PP, Lazareva MA, Savitskaya TS, Volchikhina MD (1975) J Chromatogr 108: 61
6. Tacx JCJF, German AL (1989) J Polym Sci A-1, Polym Chem 27: 817
7. Tacx JCJF, German AL (1989) Polymer 30: 918
8. Inagaki H, Tanaka T (1982) Pure Appl Chem 54: 309
9. Tanaka T, Omoto M, Donkai N, Inagaki H (1980) J Macromol Sci-Phys, B17: 211
10. Glöckner G (1989) In: Sir Geoffrey Allen (ed) Comprehensive polymer science, vol 1, Booth C, Price C (eds) Pergamon, Oxford, p 313
11. Glöckner G, van den Berg JHM, Meijerink NL, Scholte TG (1986) In: Kleintjens L and Lemstra P (eds) Integration of fundamental polymer science and technology, Elsevier Applied Science Publ, Barking, UK p 85
12. Glöckner G, Stickler M, Wunderlich W (1989) J Appl Polym Sci 37: 3147
13. Brüssau RJ, Stein DJ (1970) Angew Makromol Chem 12: 59
14. Garcia-Rubio LH (1980) Ph D Thesis, Hamilton (Canada)
15. Mori S (1988) Anal Chem 60: 1125
16. Mori S (1981) Anal Chem 53: 1813
17. Mori S (1989) (personal communication)

18. Glöckner G, Stejskal J, Kratochvíl P (1989) Makromol Chem 190: 427
19. Wild L, Ryle TR (1977) Polym Prepr, Am Chem Soc, Polym Chem Div 18: 182
20. Wild L, Ryle TR, Knobeloch DC, Peat IR (1982) J Polym Sci A-2, Polym Phys 20: 441
21. Wild L, Ryle TR, Knobeloch DC (1982) Polym Prepr, Am Chem Soc, Polym Chem Div 23: 133
22. Kelusky EC, Elston CT, Murray RE (1987) Polym Engn Sci 27: 1562
23. Usami T, Gotoh Y, Takayama S (1986) Macromolecules 19: 2722
24. Gotoh Y, Usami T, Takayama S (1988) Poster Presentation, 1st ISPAC Meeting, Toronto, June 2–3
25. Nakano S, Goto Y (1981) J Appl Polym Sci 26: 4217
26. Augenstein M, Stickler M (1990) Makromol Chem 191: 415
27. Mori S (1990) J Chromatogr 503: 411

11 Experimental Problems

11.1 THF as a Mobile Phase

A great deal of the separations reported in Chap. 9 were performed with eluents containing THF, which in gradient elution were used as component **B**. The choice of THF was often empirical but can be understood as a consequence of its favourable properties:

1. THF is a powerful solvent for many polymers. This is the reason why it is also a very popular eluent in SEC.
2. THF is miscible with apolar solvents (e.g. *iso*-octane, hexane, or cyclohexane) as well as with highly polar liquids, including methanol and water. Miscibility is prerequisite to gradient elution.
3. THF is polar enough to be used as an eluent **B** in normal-phase gradients. On the other hand, in combination with highly polar liquids as, e.g. methanol, it can be used also as component **B** in reversed-phase gradients. Its elution strength matches with a high dissolving power.
4. Since pure THF is sufficiently transparent, UV detection causes no problems when performed at e.g. 254 nm. With "sudden-transition gradients" (see Sect. 5.6) solutes can be monitored at 230 nm and, with the help of a double-beam spectrophotometer [1] detection was possible even at 215 nm.
5. THF has only moderate toxicity. The threshold limit value is 200 ppm [2] which compares favourably with tetrachloromethane (2 ppm), trichloromethane (25 ppm), dioxane or acetonitrile (40 ppm each), 1,2-dichloroethane or 1-butanol (50 ppm each), or *n*-hexane (100 ppm).

A serious drawback of THF is its tendency to form peroxides. The peroxides and their reaction products not only cause a shift of UV absorption towards higher wavelengths but also make the solvent dangerous. When overheated the peroxides decompose with extreme violence. Usually THF is stabilized by, e.g. *tert*-butyl cresol. The stabilizer contributes to UV absorbance but does not hamper SEC with THF as an eluent. In gradient HPLC pure THF without stabilizer must be used.

The stabilizer can be removed by distillation under a cover of nitrogen [3]. Before starting any operation with THF, a peroxide test is strongly recommended. Great care must be taken if the result is positive. Any treatment should be tested with small portions of the solvent. THF containing more than 0.5% peroxide should not be used at all.

THF of excellent purity was obtained from commercial THF without stabilizer by distillation over potassium in a 2-m silver-coated column and subsequent refluxing the middle fraction in a closed-circle apparatus over potassium [4]. At best, only freshly distilled THF should be used in gradient HPLC. Even with appropriate precautions, storage without stabilizer causes the absorption threshold to shift towards longer wavelengths [3].

The handling of THF must, on its way from a clean supply to the eluent reservoir of an HPLC apparatus, be well organized and swift. In containers to be filled, displacement of air by nitrogen is necessary.

Of course, the eluent must be protected against oxygen also in the HPLC apparatus. Advanced equipment provides possibilities for contineously purging the eluents. Note that not only THF but also the supplimentary gradient liquids must be protected, especially if they tend to dissolving oxygen. With clean THF under a cover of nitrogen and *i*Oct as solvent **A** without precautions, a blank gradient (5–90% **B** in 15 min) yielded a nonlinear transition from the initial to the final signal level with a pronounced maximum at about 50% **B**. The oxygen dissolved by *i*Oct was transferred to THF yielding immediately UV absorption [3]. Purging also eluent **A** with nitrogen caused, in repeated cycles of blank gradients, the maximum to diminish; after about two hours a linear transition could be observed.

Another disadvantage of THF is its hygroscopicity. Water strongly influences adsorption activity of columns and is furthermore an extremely powerful precipitant for organophilic polymers. Even traces of water influence the properties of polymer solutions dramatically [5]. Thus, THF must be protected against oxygen as well as against moisture. The latter also holds true with stabilized THF and must not be ignored in SEC when this eluent is used [6].

11.2 Sample Solvent

Active columns support the proper retention of polymer samples. Under favourable conditions, the starting eluent can be used as solvent for the polymer to be chromatographed. In case of improper retention, the starting eluent should be a nonsolvent for the polymer [7] and, hence, the sample must be dissolved in a liquid of higher thermodynamic quality.

In precipitation HPLC, the choice of the sample solvent is not arbitrary. The solvent should be identical with a gradient component, otherwise it would act as an extra sample. In polymer HPLC, the injection volume is about 10 μl which, at a density of 1 g/cm^3, corresponds to 10,000 μg. If not a gradient component, this low-MW extra sample would strongly overload the column, and the protracted elution of this solute would disturb the separation of the polymer sample. (In isocratic RP chromatography, peak distortion has been observed as a consequence of sample solvents stronger than the mobile phase [8]). If the sample solvent is a gradient component, an excess volume of 10 or even 25 μl is easily digested by a chromatographic system.

In this respect, combinations of sample solutions in THF with gradients employing THF as component **B** (or with "sudden-transition gradients" at a 20–30% level of THF content) are proper systems. They are especially suited for CCF with prefractionation by SEC in THF eluent.

One need not worry about a stabilizer in the sample solvent. THF is usually stabilized by the addition of 0.025% of a suitable compound. This concentration corresponds, at $V_0 = 10\text{–}100\ \mu l$, to an extra sample with $m_0 = 2.5 - 25\ \mu g$ which is well within the load capacity of a column. Provided that the elution volume does not coincide with that of a polymer component, the stabilizer will not disturb the chromatography of polymer samples.

Even when the sample solvent is a gradient component, the solvent plug is usually visible in an optical detector. This is basically due to different absorption, but may be additionally affected by a difference in refractive index, see Sects. 7.1.1 and 11.3.

11.3 Ghost Peaks

In earlier investigations with an UV detector, pseudo-peaks were observed even at changes in the slope of multilinear gradients. In order to suppress these artefacts which were due to a difference in refractive index between eluent **A** (*iso*-octane, $n_D = 1.392$) and THF ($n_D = 1.405$), a mixture of the latter with about 10% methanol ($n_D = 1.329$) was tried as an eluent **B**. Since polymer elution was not disturbed and pseudo-peak formation indeed suppressed, the eluent system mentioned was repeatedly employed in investigations of S/AN copolymers.

With present knowledge it must be appreciated that MeOH addition facilitates the separation of S/AN by composition. This was found in recent investigations [9] with a tapered detector cell which is almost insensitive to changes in refractive index.

Ghost peaks occurred on polar columns with some brands of *i*Oct used as gradient component **A**, where obviously traces of impurities were trapped on the column during the prerun period. These unknown components were eluted in narrow peaks with reproducible retention times, due to displacement by eluent mixtures of adequate polarity. The origin of these peaks became evident by the linear variation of their size with the length of the prerun period. The addition of 2% MeOH throughout suppressed the ghost peaks [7].

11.4 Sample Load and Column Blocking

In principle, injection into a nonsolvent can cause column blocking but, in thousands of injections, we never have been confronted with this accident, provided that the sample solution was homogeneous. With respect to the latter, difficulties can arise with mixed calibration standards, see Sect. 11.5.

The likelihood of blocking, which depends on the system to be investigated and on the grain size of the packings, can be diminished by reducing the column load. With column packings of 5 μm particle diameter, we applied at maximum

62.7 μg S/AN ($V_0 = 30\,\mu$l) or 32 μg S/MMA ($V_0 = 20\,\mu$l) on a 150 × 4.6 mm silica column. The starting eluent was *i*Oct/THF (90:10) in these cases. At maximum, 400 μg PS (MW 110,000) or 200 μg PS (MW 470,000) in $V_0 = 10\,\mu$l could be injected into MeOH/DCM (80:20) on a C18 column 250 × 4.1 mm [10]. Mori reported injection of 120 μg S/MMA in $V_0 = 100\,\mu$l eluent **A** on a 50 × 4 mm column [11a] or even 800 μg S/VAC block copolymer in $V_0 = 200\,\mu$l on a 50 × 4.6 mm column [11b].

The likelihood of blocking increases with decreasing grain size. In a paper dealing with gradient HPLC of epoxy composite formulations, a warning has been given not to inject more than 6 μl sample solution containing 36 μg prepolymer in THF on a 40 × 4.6 mm C18 column packed with 3 μm particles. Packings with 5 μm particles were found to be less sensitive to sample size [12].

Assume an injection volume of 10 μl containing 10 μg polymer in THF and injection into a starting eluent *i*Oct + THF (90:10). Assume also the total pore volume to be equal to the interstitial volume of the column, and the distribution coefficient of THF to be equal to one. Then the eluent available in a pore volume of about 20 μl should suffice to reduce the THF content in the sample volume to 40% where a S/AN copolymer is no longer kept in solution. With spheres of 10 μm diameter and a porosity of $\varepsilon_P = 0.4$, the required pore volume of 20 μl corresponds to an amount of packing material with an outer surface of 300 cm^2. Thus, the polymer will be deposited at a density of 0.01 mg/0.03 m^2 or 0.33 mg/m^2. This is only one third of the value typical of a monomolecular layer in polymer adsorption, 1 mg/m^2. The thickness of a film corresponding to 1 mg/m^2 is less than 0.1 μm. Although this simplified estimation ignores a gradual alteration in solvent composition, it can explain why blocking was not observed.

Usually, about 10 μg were injected on a column of 4.6 mm inner diameter. With a sample load of this order, i.e. about 0.6 μg per mm^2 column cross-section, one is certainly on the safe side. At any rate, the behaviour of unknown systems should be explored cautiously, starting with injections as small as possible. Observation of the column back-pressure during pioneering runs is recommended. For example, see the pressure diagram for the elution of S/AN through a Hex/THF gradient in Ref. [13]. It indicated that a pressure increase during a gradient cycle was due to only the higher viscosity of THF as compared with hexane but not to a contribution of sample polymer.

11.5 Hazards Caused by Incompatible Sample Polymers

Incompatibility is a wide-spread phenomenon in polymer behaviour which, e.g. can cause precipitation of polymeric material from mixed solutions. This risk must be taken into account if calibration is carried out using mixtures of standard copolymers graded in composition. Even copolymers of one and the same monomeric system can be incompatible and thus, precipitate from mixed solutions. Well-known examples are S/AN copolymers where a 5% difference in composition suffices for incompatibility [14].

The problem lies especially in a kinetic hindrance which may allow a metastable mixture to remain clear for hours or days, but can cause precipitation at any unforeseen moment. Thus, frequent inspection of mixed sample solutions is strongly recommended in order to avoid blocking the injection system with precipitated polymer.

11.6 Column Flushing

In order to restore the initial activity of a column thorough flushing after each run is essential. This general recommendation is of special importance in polymer chromatography where some portions of the sample may be not completely eluted during a gradient program. These traces would lead to drifting retention data, memory effects (i.e. elution in subsequent gradient cycles or blank runs), and reduced column life.

The effect of insufficient column regeneration on retention data can be seen in Fig. 11.1 for tetraglycidyl-4,4′-diamino diphenylmethane, a main component of an epoxy composite [12].

It is good practice to use return gradients for restoring column activity. The return gradients should include liquids which are on the one hand good solvents for the polymers investigated as well as chromatographically powerful eluents. THF is quite suitable. Rinsing with a tenfold column volume of pure THF has proved success [15].

Very good results were obtained with cleaning cycles from 100% THF to 100% MeOH and back to 100% THF, followed by a cycle to 100% *i*Oct and back

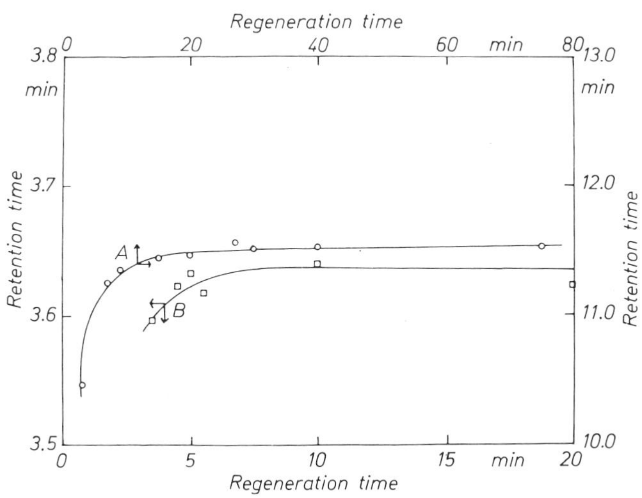

Fig. 11.1. Retention time of an epoxy compound in RP gradient elution as a function of regeneration time. A: C18-column 300 × 4 mm, $d_p = 5\,\mu m$, elution by water/acetonitrile (50–100% in 15 min), return gradient W/AcN 100–50% in 1 min, flow rate 1 ml/min. B: C18-column 40 × 4.6 mm, $d_p = 3\,\mu m$, elution by water/acetonitrile (20–100% in 3 min), return gradient W/AcN 100–200% in 0.1 min, flow rate 2.0 ml/min. Sample: tetraglycidyl-4, 4′-diamino diphenylmethane (peak #4 in Fig. 9.31), HPLC at 25 °C. From Ref. [12] with permission

to 100% THF before returning to the initial eluent composition. The period of this procedure could be shortened by increasing the flow rate to 2 ml/min. (The analytical flow rate was 0.5 ml/min.)

11.7 Effect of Column Length

In gradient HPLC of proteins, separation efficiency was as high on short columns as on longer ones, or even higher. Remarkable results were, e.g. obtained with a test mixture of ribonuclease A, insulin, cytochrome C, bovine serum albumin, and ovalbumin on RP C18 columns. The resolution gained with a column 6.3 mm (!) in length and 4.6 mm in diameter was more than twice as high as the resolution on a 45 × 4.6 mm column of otherwise identical properties, provided the flow rate was 0.5 or 0.25 ml/min and the volumetric gradient-rate large (16% **B** per ml). Eluent **A** and **B** were water and isopropanol, respectively, both with 0.1% trifluoro acetic acid [16].

With S/AN copolymers, a better separation was obtained on a 55 × 4.6 mm column than on a 150 × 4.6 mm one [17]. All other conditions including packing material (silica Polygosil 60–5, $d_P = 5\,\mu m$, $d_0 = 6\,nm$), gradient (*i*Oct/THF, 5%/min), and flow rate (1 ml/min) were identical. The diffusion of eluent components, which is definitely faster than the diffusion of the high-MW sample, can obviously cause band broadening. This effect will increase with increasing gradient steepness, decreasing flow rate, and increasing column length.

The drawback of short columns is their low sample capacity. In Sect. 8.3 difficulties with short columns in combination with large injection volumes are reported.

11.8 Peak Splitting

Multimodal peaks and oscillating UV signals of broad elution patterns have been observed in the elution of synthetic polymers [18, 19] or nucleic acids [20, 21]. Although reproducible on a given apparatus [18], they are artefacts caused by extremely small oscillations in eluent composition. The latter are scarcely noticable in low-MW HPLC or in the elution of uniform biopolymers where narrow peaks occur. There they might, at worst, cause some shift in retention time, but the broad elution patterns of polymers are susceptible to oscillations in eluent composition. Because of the high values of *solute acceleration factor S*, tiny changes in eluent composition markedly affect the elution strength of a mixed eluent. Thus, patterns as reported in Ref. [18] occur, especially when detection with a low attenuation is needed. Needless to say that these artefacts must be distinguished from multiple peaks due to unfolding of proteins [22].

11.9 Column Conditioning by Sample Injection

In investigations of new stationary phases, the occurrence of a "condition period" was sometimes observed. The term refers to the first injections on a freshly made column where the peak height increases to a constant value with repeated injections. This was found, e.g. in hydrophobic interaction chromatography of proteins [23].

A similar effect was observed by injections of 8.2 μg of the mixture of five S/AN copolymers in $V_0 = 20\,\mu$l THF on a brandnew column packed with ground glass [17]. The tracing of the first injection was smaller than those from immediate repetitions. With a forth injection, the condition period was over.

Another but possibly related observation was reported in 1984 [24]: S/AN samples of broad distribution (precisely, commercial copolymers without reprecipitation etc.) yielded on the first injection after an extended flushing period a higher peak than in a subsequent repetition. A following third and forth injection caused almost no further change of the tracing. Unfortunately, it still holds that "the cause of this erratic behaviour, which never occurred with the prefractionated samples" is not known. Nevertheless, the strange observation was a challenge to modify chromatographic condition towards straightforward systems without any extra components and with strictly linear gradient programs.

11.10 Guidelines for Selecting Suitable Phase Systems for Copolymer HPLC

The first question concerning the mobile phase refers to the equipment available. With an evaporative light-scattering detector, the optical properties of the liquids are not critical. When the elution is to be monitored by a UV detector, all eluent components must be transparent at the chosen wavelenth. (A possibility of escaping this limitation is described in Sect. 5.6.)

The columns used in a search should measure about 150 × 5 mm I.D. and be packed with small pore packings ($d_0 \leqq 5$ nm, $d_P < 10\,\mu$m). Then selection can proceed by the following steps.

1. Find a solvent which, by boiling point, toxicity, price etc. is suited for HPLC and capable of dissolving the copolymers to be investigated in the whole range of composition and MW of interest.
2. If the copolymer is composed of monomeric units sufficiently differing in polarity (e.g. S/AN or S/MMA) look for a nonpolar liquid which is miscible with the solvent selected.
3. Run the nonpolar liquid as eluent **A** on a CN bonded phase and inject a small amount of copolymer ($m_0 \leqq 10\,\mu$g). Watch the column inlet pressure. The sample must be retained. If retention is not gained, try a nonpolar precipitant as eluent **A**, or use a silica column, or both.
4. If the sample is properly retained, apply a survey gradient (e.g. 0–100% **B** in 10 min) and watch sample elution. If elution fails be cautious with repeated injections. A less polar column or the admixture of a strongly polar liquid to eluent **B** might help.
5. If all attempts fail or elution patterns are badly shaped go back to step No. 2 and look for a polar nonsolvent which is miscible with the solvent.

6. Run this polar nonsolvent as eluent **A** on a C18 column and inject. Watch the column inlet pressure. If retention fails, increase the polarity of the RP eluent **A** by addition of a nonsolvent of higher polarity. The ultimate choice may be even the addition of water in small portions.
7. On proper retention, elute the sample by a survey gradient. If elution fails try another solvent (or mixed solvents) as an eluent **B**.
8. If retention and elution have been achieved either in the normal-phase mode (steps 2–4) or in the RP mode (steps 5–7) repeat the respective steps with a copolymer of different composition and look for a reasonable change in elution time. If this expectation is not fulfilled with the specimens used, try more samples. If all attempts fail try the opposite mode (e.g. RP instead of NP conditions). The last hope will be the search for a more suitable solvent. Repeat from step No. 1.
9. If separation by composition is reached, investigatee the elution quantitatively. Is the polymer properly retained in the range of sample size needed for eventual analyses? Is the polymer quantitatively eluted? Does sample size affect retention time?
10. Optimize gradient and flow rate to the problem under investigation and calibrate.

11.11 References

1. Johnson EL, Gloor R, Majors RE (1978) J Chromatogr 149: 571
2. Threshold Limit Values, annual list, American Conference of Governmental Industrial Hygenists, Boston USA
3. Glöckner G, van den Berg JHM (1984) Chromatographia 19: 55
4. Müller AHE personal communication
5. Spychai T, Lath D, Berek D (1979) Polymer 20: 437
6. Spychai T, Berek D (1979) Polymer 20: 1108
7. Glöckner G (1987) Chromatographia 23: 517
8. Hoffman NE, Pan S-L, Rustum AM (1989) J Chromatogr 465: 189
9. Schultz R, Engelhardt H (1990) Chromatographia 29: 325
10. Glöckner G, Schmutzler S, Engelhardt H, Schultz R (1988) Chromatographia 25: 983
11a. Mori S (1988) Anal Chem 60: 1125
11b. Mori S (1990) J Chromatogr 503: 411
12. Noel D, Cole KC, Hechler J-J, Chouliotis A, Overbury KC (1986) J Appl Poly Sci 32: 3097
13. Glöckner G, Kroschwitz H, Meissner Ch (1982) Acta Polymerica 33: 614
14. Molau GE (1965) J Polym Sci B, Polym Letters 3: 1007
15. Danielewicz M, Kubin M (1981) J Appl Polym Sci 26: 951
16. Moore RM, Walters RR (1984) J Chromatogr 317: 119
17. Glöckner G, van den Berg JHM (1987) Chromatographia 24: 233
18. Glöckner G (1988) Chromatographia 25: 854
19. Augenstein M (1989) (personal communication)
20. Garcia S, Liautard J-P (1983) J Chromatogr Sci 21: 398
21. Garcia S, Liautard J-P (1984) J Chromatogr 296: 355
22. Benedek K, Dong S, Karger BL, (1984) J Chromatogr 317: 227
23. Miller NT, Feibush B, Karger BL (1984) J Chromatogr 316: 519
24. Glöckner G, van den Berg JHM, Meijerink NLJ, Scholte TG, Koningsveld R (1984) J Chromatogr 317: 615

12 Glossary of Terms

Alternating copolymer
Apparent molecular weight
Azeotropic copolymer
BAKER–WILLIAMS fractionation
Bipolymer
Block copolymer
Capacity factor
Copolymer composition: mass composition
Copolymer composition: molar composition
Copolymer mixtures
Copolymer mixtures: numerical examples
Copolymerization diagram
Copolymerization propagation probab.
Dead time
Degree of conversion
Degree of polymerization
Dwell time
Gamma function
Gel effect
Gradient elution
Graft copolymer
Height of a theoretical plate
Interstitial volume
Intrinsic viscosity
Isocratic elution
Mayo–Lewis equation
Molecular weight
Monomer reactivity ratios
Most-probable chain-length distrib.
Net retention time
Normal-phase chromatography
Number average
Pore volume
Random copolymer
Recursive Mayo–Lewis equation
Reversed-phase chromatography
Schulz–Zimm distribution
Solute acceleration factor
Statistical copolymer
Terpolymer
θ-System
van Deemter equation
Weight average

Alternating Copolymer
Bipolymer whose macromolecules contain the constituting units in equimolar amount and in a regularly alternating sequence.

Apparent Molecular Weight of a Repeat Unit, M_{app}
Homopolymers are composed of uniform repeat units whose M_0 value can be calculated by the sum of the atomic weights of the constituting elements. M_{app} corresponds to M_0 and is useful for calculating the *degree of polymerization* of

copolymers. It can be calculated from the M_0 data of the monomeric units, e.g. $M_A, M_B, M_C, \ldots$, for the constituents A, B, C, etc. and the *molar composition* of the copolymer:

$$M_{app} = x_A M_A + x_B M_B + x_C M_C + \cdots \tag{12.1}$$

For bipolymers, this equation reduces to:

$$M_{app} = M_A + x_B(M_B - M_A) \tag{12.2}$$

Azeotropic Copolymer
Copolymer obtained from an azeotropic monomer mixture. It has, independent of the degree of conversion, the same composition as the monomer mixture. The condition for binary azeotropic monomer mixtures can be derived from the *Mayo–Lewis equation:*

$$[A]_0/[B]_0 = (r_B - 1)/(r_A - 1) \tag{12.3}$$

In a *copolymerization diagram,* the azeotropic point is the intersection of the copolymerization curve with the straight line determined by $r_A = r_B = 1$. Thus, only systems with $r_A < 1$ and $r_B < 1$ (or $r_A > 1$ and $r_B > 1$) have an azeotropic point.

BAKER–WILLIAMS Fractionation
Solubility-based method of polymer separation by the combined effects of a temperature gradient along the column and gradient elution. The latter is performed with solvent/nonsolvent mixtures whose thermodynamic quality increases in the course of the run. The column is packed with inert nonporous material, e.g. glass beads. The samples are, outside the column, deposited on a portion of the packing material which is then added as the top part of the column packing ("sample bed"). The basic idea of BAKER–WILLIAMS fractionation embraces the dissolution of a sample portion at a higher temperature and its subsequent precipitation at a lower temperature after a short period of migration. At its new position, the sample portion is redissolved by an eluent mixture of higher dissolving power. Multiple repetitions of precipitation and redissolution ensure a high separating efficiency of the method.

Baker CA, Williams RPJ (1956) J Chem Soc (London) 2352

Bipolymer (also: **binary copolymer**)
Copolymer produced from two species of monomers.

Block Copolymer
Copolymer composed of extended sequences, each containing only one monomeric species. The number of blocks is usually known, e.g. a **diblock copolymer** of A and B units is a polymer with an uninterrupted sequence of A units linked to a corresponding block of B units.

Capacity Factor k'
Ratio of the *net retention time* t'' of a solute and the column *dead time* t':

$$k' = t''/t' \tag{12.4}$$

The capacity factor allows a straightforward comparison of chromatographic results obtained with different apparatus.

Copolymer Composition: Mass Composition
Measure of copolymer composition by using the weight fractions w_A, w_B, $w_C, \ldots$ of the constituents A, B, C, ...:

$$w_A = m_A/(m_A + m_B + m_C + \cdots) \tag{12.5}$$

The values m_A, m_B, $m_C, \ldots$ represent the mass of the respective monomeric units in a copolymer. The weight fraction w_i is related to mole fractions x_i by:

$$w_A = x_A M_A/(x_A M_A + x_B M_B + x_C M_C + \cdots) \tag{12.6}$$

For bipolymers and $M_A/M_B = q$, Eq. (12.6) reads:

$$w_A = \frac{x_A q}{x_A q + (1 - x_A)} \tag{12.7}$$

or

$$w_B = \frac{x_B}{(1 - x_B)q + x_B} \tag{12.8}$$

In the present pages, the composition of bipolymers is given by the content in the last-mentioned monomer, e.g. "copoly(styrene/butadiene), 10.2 mass%" means 10.2% butadiene. (With bipolymers, the full description "copoly(S/Bd), (89.8: 10.2 mass%)" would be redundant.)

Copolymer Composition: Molar Composition
Measure of copolymer composition by using the mole fractions x_A, x_B, $x_C, \ldots$ of the monomeric units A, B, C, etc., in a copolymer. The mole fractions have the same meaning as in mixed phases:

$$x_A = n_A/(n_A + n_B + n_C + \cdots) \tag{12.9}$$

Mole fractions can be calculated from weight fractions, w, by using $n_i = m_i/M_i$ etc. and $w_i = m_i/\sum m_i$, through

$$x_A = \frac{w_A/M_A}{w_A/M_A + w_B/M_B + w_C/M_C + \cdots} \tag{12.10}$$

In Eq. (12.10), w_i is substituted for m_i since $\sum m_i$ is contained in the dividend as well as in the divisor. For bipolymers and $M_A/M_B = q$, Eq. (12.10) reads:

$$x_A = \frac{w_A}{w_A + (1 - w_A)q} \tag{12.11}$$

or

$$x_B = \frac{w_B q}{(1 - w_B) + w_B q} \tag{12.12}$$

Values of M_A and M_B are given in Tables 9.1–9.10.

For copolymers composed of monomeric units with approximately equal M_0, e.g. styrene ($M_0 = 104.1$) and methyl methacrylate ($M_0 = 100.1$), the differences between the numerical values of mole fractions and weight fractions are almost within the limits of experimental error. Attention is required with copolymers composed of units with substantially differing M_0, see *copolymer mixtures: numerical examples.*

Copolymer Mixtures

Model mixtures of copolymers are essential for development and testing of separation methods. This section deals with the average concentration of the constitutent B in mixtures of copolymers from a given binary system.

Mixtures are usually composed by weighing. Assume the mass of bipolymers *1, 2, 3,*... in the mixture be $f_1, f_2, f_3, \ldots$ and the respective weight fraction of monomer B in the bipolymers $w_1, w_2, w_3, \ldots$, than the average weight fraction of constituent B in the mixture is:

$$w_{av} = \frac{w_1 \cdot f_1 + w_2 \cdot f_2 + w_3 \cdot f_3 + \cdots}{f_1 + f_2 + f_3 + \cdots} \tag{12.13}$$

The average mole fraction, x_{av}, of the constituent B in this mixture is

$$x_{av} = \frac{x_1 \cdot g_1 + x_2 \cdot g_2 + x_3 \cdot g_3 + \cdots}{g_1 + g_2 + g_3 + \cdots} \tag{12.14}$$

where $g_1, g_2, g_3, \ldots$ are the molar portions of bipolymers *1, 2, 3,*..., which contain mole fractions $x_1, x_2, x_3, \ldots$ of B units.

The average mole fraction can be calculated also from w_{av} by Eq. (12.12) using $w_B = w_{av}$. This can be understood by a reasoning as follows: Since $w_i f_i = m_i$, it follows from Eq. (12.13) for the total mass of A or B units in the mixture

$$\sum m_{A,i} = \sum (1 - w_i) f_i = (1 - w_{av}) \sum f_i \tag{12.15}$$

and

$$\sum m_{B,i} = \sum w_i \cdot f_i = w_{av} \sum f_i \tag{12.16}$$

The relations $n_{A,i} = m_{A,i}/M_A$ and $n_{B,i} = m_{B,i}/M_B$ also hold true for the whole mixture:

$$\sum n_{A,i} = \frac{\sum m_{A,i}}{M_A} = \frac{(1 - w_{av}) \sum f_i}{M_A} \tag{12.17}$$

and

$$\sum n_{B,i} = \frac{\sum m_{B,i}}{M_B} = \frac{w_{av} \sum f_i}{M_B} \tag{12.18}$$

Eq. (12.14) can, with Eqs. (12.17 and 18) and $q = M_A/M_B$, be rewritten as

$$x_{av} = \frac{\sum n_{B,i}}{\sum n_{A,i} + \sum n_{B,i}} = \frac{w_{av} \cdot q}{(1 - w_{av}) + w_{av} \cdot q} \tag{12.19}$$

which indeed is equivalent to Eq. (12.12).

The individual molar concentration g_i of a copolymer in the mixture is

$$g_i = f_i \left(\frac{1 - w_{B,i}}{M_A} + \frac{w_{B,i}}{M_B} \right) \tag{12.20}$$

Eq. (12.20) is based upon the relation for the mole fraction $x_i = n_i/g_i$ (with $g_i = n_{A,i} + n_{B,i}$) and molar ("$n_{B,i}$") or mass contributions ("$m_{B,i}$") of B units due to copolymer *i*. The relation

$$g_i = \frac{n_{B,i}}{x_i} = \frac{m_{B,i}/M_B}{x_i} \tag{12.21}$$

gives, by substituting Eq. (12.12) for x_i, Eq. (12.20).

The *apparent molecular weight of a repeat unit* of copolymer *i* is

$$M_{app,i} = \frac{f_i}{g_i} = \left(\frac{1 - w_{B,i}}{M_A} + \frac{w_{B.i}}{M_B} \right)^{-1} \tag{12.22}$$

Copolymer Mixtures: Numerical Examples

The difference between mass fraction and mole fraction is large if the ratio M_A/M_B differs significantly from the value 1. The latter holds true, e.g. for copolymers of styrene ($M_A = 104.144$) and butadiene ($M_B = 54.088$) where $M_A/M_B \approx 2$. (Precisely, $M_A/M_B = 1.925$, but for convenience the approximation will be used.)

Example I, Weight Composition of Copolymers Given:

Assume a copolymer *1* with $w_{B,1} = 0.1$ and another one (*2*) with $w_{B2} = 0.9$. An 1:1 mixture (w/w) of both specimens has an average composition of $w_{av} = (0.1 \times 1 + 0.9 \times 1)/2 = 0.5$ in monomeric unit B. The corresponding mole fractions are, according to Eq. (12.12), (with $M_A/M_B = 2$):

$$x_{B1} = [0.1 \times 2]/[(1 - 0.1) + 0.1 \times 2] = 0.182$$

$$x_{B2} = [0.9 \times 2]/[(1 - 0.9) + 0.9 \times 2] = 0.947$$

and, according to Eq. (12.19),

$$x_{av} = [0.5 \times 2]/[(1 - 0.5) + 0.5 \times 2] = 0.667$$

Note that $x_{av} = (0.182 + 0.947)/2 = 0.565$ would be valid for an equimolar mixture but not for the combination of equal-mass portions.

Example II, Molar Composition of Copolymers Given:
Assume a copolymer *1* with $x_{B1} = 0.1$ and another one (*2*) with $x_{B2} = 0.9$. The corresponding weight fractions are, according to Eq. (12.8):

$$w_{B1} = 0.1/[(1 - 0.1) \times 2 + 0.1] = 0.053$$
$$w_{B2} = 0.9/[(1 - 0.9) \times 2 + 0.9] = 0.818$$

An equal-weight mixture of both components has, according to Eq. (12.13)

$$w_{av} = (0.053 + 0.818)/2 = 0.436$$

which can, by Eq. (12.19), be used for calculating the average mole fraction of monomeric units B:

$$x_{av} = [0.436 \times 2]/[(1 - 0.436) + 0.436 \times 2] = 0.607$$

This demonstrates that the molar composition of a mixtures of mass portions must not be calculated by averaging mole fractions. The mean value $x_{av} = (0.1 + 0.9)/2 = 0.5$ is valid for an equimolar mixture, vide supra.

Example III, Mixture of four Bipolymers:
Assume *bipolymers* of monomer A and monomer B with $M_A = 200$ and $M_B = 100$ and a mixture composed of 0.5 g of specimen "*1*" having $w_{B,1} = 0.2$, 0.5 g of "2" ($w_{B,2} = 0.4$), 0.6 g of "*3*" ($w_{B,3} = 0.5$), and 1.0 g of "*4*" ($w_{B,4} = 0.8$). Table 12.1 compiles data of this mixture calculated from these values. The averages are $w_{av} = 1.4/2.6 = 0.538$ (according to Eq. (12.13)) and $x_{av} = 0.014/0.02 = 0.7$ (according to Eq. (12.14)).

Copolymerization Diagram
Graphical representation of the *Mayo–Lewis equation*: plot of the mole fraction of a monomeric unit in a copolymer vs the respective mole fraction in the starting monomer mixture. The **copolymerization curve** is a straight line when the *monomer reactivity ratios* have the values $r_A = r_B = 1$. With $r_A, r_B < 1$ (see Fig. 2.4), an intersection of the copolymerization curve and the diagonal straight line exists

Table 12.1. Average composition of a mixture of four bipolymers

No.	Weight composition $w_{B,i}$	Amount f_i(g)	Amount of B monomer $m_{B,i}$(g)	Number of moles $10^3\,n_{B,i}$	Mole fraction $x_{B,i}$	n_i/x_i $10^3\,g_{B,i}$
1	0.2	0.5	0.1	1	0.333	3
2	0.4	0.5	0.2	2	0.571	3.5
3	0.5	0.6	0.3	3	0.667	4.5
4	0.8	1.0	0.8	8	0.889	9
	Total	2.6	1.4	14		20

which indicates an *azeotropic copolymer*. Systems with $r_A > 1$ and $r_B < 1$ have no azeotropic point.

Copolymerization Propagation Probabilities

The likelihood of the addition of a monomer A to a polymerizing molecule whose active site is formed by a previously added unit A is given by

$$p_{AA} = \frac{k_{AA}[A]}{k_{AA}[A] + k_{AB}[B]} \tag{12.23}$$

where k_{AA} and k_{AB} are the reaction rate constants for the addition to an ultimate A unit, and [A] and [B] the molar concentrations of A and B in the monomer mixture, respectively. Equivalent expression can be written for the addition of B to A, B to B, and A to B:

$$p_{AB} = \frac{k_{AB}[B]}{k_{AA}[A] + k_{AB}[B]} \tag{12.24}$$

$$p_{BB} = \frac{k_{BB}[B]}{k_{BA}[A] + k_{BB}[B]} \tag{12.25}$$

$$p_{BA} = \frac{k_{BA}[A]}{k_{BA}[A] + k_{BB}[B]} \tag{12.26}$$

The Eqs. (12.23) to (12.26) can be rewritten by using *monomer reactivity ratios*, r_A and r_B, together with the molar concentration ratio $G = [A]/[B]$ or the mole fraction in the monomer batch, $X = X_A = [A]/([A] + [B])$. For example, Eq. (12.23) gives:

$$p_{AA} = \frac{r_A G}{1 + r_A G} = \frac{r_A X}{1 + r_A X - X} \tag{12.27}$$

Dead Time, t'

Elution time of an unretained solute. In interactive HPLC with unrestricted accessibility of pores, e.g. adsorption chromatography, t' is equal to the elution time of the mobile-phase volume of a column (i.e. *pore volume* plus *interstitial volume*), whereas in exclusion chromatography or interactive HPLC of large-size solutes on small-pore packings, t' is equal to the elution time of the liquid in the interstitial volume.

Degree of Conversion, ψ

The degree of molar conversion in a binary copolymerisation is

$$\psi = \frac{([A]_0 - [A]) + ([B]_0 - [B])}{[A]_0 + [B]_0} \tag{12.28}$$

With the average composition of the copolymer

$$\bar{x}_A = \frac{[A]_0 - [A]}{([A]_0 - [A]) + ([B]_0 - [B]} \tag{12.29}$$

and the composition on the monomeric mixture $X_A = [A]/([A] + [B])$ and $X_{A,0} = [A]_0/([A]_0 + [B]_0)$ at degree of conversion ψ or zero, respectively, the relation holds

$$\bar{x}_A = \frac{X_{A,0} - (1 - \psi)X_A}{\psi} \tag{12.30}$$

Degree of Polymerization, *P*
Number of repeat units in a macromolecule. The degree of polymerization can be calculated from the *molecular weight* of a polymer, M, and the molecular weight of a repeat unit, M_0. This is usually done by

$$M = P \cdot M_0 \tag{12.31}$$

which ignores the weight of the endgroups. This is correct only with macrocycles or polymers whose M approaches infinity. For polymers of normal MW, endgroup contributions are fortunately so small that they can be neglected also in this case. With low-MW polymers they must be considered.

The meaning of Eq. (12.31) is straightforward for homopolymers which indeed consist of repeat units of uniform M_0. With copolymers, certain properties are better understood on base of chain length than of molecular weight. Thus, Eq. (12.31) also has significance for copolymers, where the *apparent molecular weight of a repeat unit*, M_{app}, must be substituted for M_0.

Dwell Time, t_{lag}
A gradient program list events which control the operation of pumps or mixing devices. The dwell time is the delay of detector signal in response to gradient program. Like *dead time* t', t_{lag} varies inversely with flow rate; t_{lag} is longer than t' because it additionally includes the period required for the flow of eluent from the mixing device to the position where samples are injected (column inlet). The dwell time can be evaluated by applying an eluent composition step and measuring the half-height position of the transition in detector signal.

Gamma Function

$$\Gamma(\mathrm{a} + 1) = \int_0^\infty \exp(-\mathrm{t}) \cdot \mathrm{t}^\mathrm{a}\,\mathrm{dt} \tag{12.32}$$

which, for positive integer a, has the value of a faculty, $\mathrm{a}! = \Gamma(\mathrm{a} + 1)$.

Gel Effect
The increase of viscosity during polymerization hampers, in the first instance, the termination reaction. This yields macromolecules of higher MW. A gel effect is to be considered at a high degree of conversion.

Gradient Elution (antonym: *isocratic elution*)
Elution by mixed eluents whose content in a stronger component increases in the course of a run. The more powerful eluent "**B**" reduces the *capacity factor* of

retained sample portions so that the chromatogram of broad mixtures becomes shorter. The effect of a gradient on retention can be indicated by the *solute acceleration factor S*. A gradient starting, e.g. with a mixture of 80% hexane and 20% tetrahydrofuran and reaching 100% THF after 20 min is indicated as Hex/THF (20–100% in 20 min). The more powerful eluent is always mentioned in the second position; concentrations are given in vol% and linear increase of **B** is assumed (exceptions are indicated, e.g. by suffix "multilinear").

Graft Copolymer
Branched polymer whose side chains are chemically different from the backbone.

Height of a Theoretical Plate, *h*
Measure of the kinetics of mass transport; h has the dimension of a length and can be calculated by subdivision of the length L of a chromatographic path by the **number of theoretical plates**, N. The latter is the squared ratio of peak elution time t_e and the standard deviation σ of a Gaussian peak (half width at 60.7% maximum height):

$$h = L/N = \frac{L}{(t_e/\sigma)^2} \tag{12.33}$$

Interstitial Volume
Total volume in a packed column which is not occupied by the particles of the packing. The closest spherical packing has a bulk factor of 74.05%, hence, the interstitial volume is, in this case, 25.95% of the column volume. Since proper packings do not reach the density of a closest packing, the interstitial volume is usually larger than 26%. With small- or non-porous packings, macromolecules are restricted to the interstitial volume.

Intrinsic Viscosity, [η]
Limiting value of the ratio η_{sp}/c at concentration $c = 0$ and zero shear rate. $[\eta]$ can be measured by the ratio η_{sp}/c and extrapolation to zero concentration after plotting vs. concentration or η_{sp}, according to Huggins or Schulz–Blaschke, respectively. $[\eta]$ has the dimension of a specific volume and is, in the present pages, given in ml/g.

Isocratic Elution (antonym: *gradient elution*)
Liquid chromatography with a pure liquid or a mixture of constant composition as a mobile phase. The composition of an isocratic mixture of, e.g. one volume of toluene and two volumes of hexane is indicated as "Tol + Hex (1:2)".

Mayo–Lewis Equation (see also: *recursive Mayo–Lewis equation*)
Copolymerization equation, relating the composition of a copolymer to the composition of the monomer mixture where it is polymerized from.

Kinetic analysis of the statistical copolymerization of A and B monomers yields:

$$\frac{n_A}{n_B} = \frac{[A]}{[B]} \cdot \frac{r_A[A] + [B]}{[A] + r_B[B]} \tag{12.34}$$

where n_A and n_B are the molar concentrations of A and B units in the copolymer, while [A] and [B] indicate the respective concentrations in the monomer batch; r_A and r_B are the *monomer reactivity ratios.* The derivation of Eq. (12.34) starts from the reaction rate expressions of the four possible monomer additions (v_{AA}, v_{AB}, v_{BB}, v_{BA}) and is based upon the rate balance

$$\frac{-\mathrm{d}[A]}{-\mathrm{d}[B]} = \frac{v_{AA} + v_{BA}}{v_{BB} + v_{AB}} \tag{12.35}$$

The assumptions in deriving Eq. (12.34) are: (i) equivalent mechanism of the addition of A or B monomers (e.g. free radical reaction, anionic reaction, etc.), (ii) homogeneous reaction, (iii) high degree of polymerization (i.e. negligible monomer consumption in reactions other than chain growing), (iv) only ultimate units influencing the addition of next monomer, and (v) irreversibility of each monomer addition.

With $g = n_A/n_B$ and $G = [A]/[B]$, Eq. (12.34) can be rewritten as:

$$g = G\frac{r_A G + 1}{G + r_B} \tag{12.36}$$

Similarly, with the mole fractions $x = x_A = n_A/(n_A + n_B)$ and $X = X_A = [A]/([A] + [B])$ in the copolymer and the monomeric mixture, respectively, Eq. (12.34) becomes

$$x = \frac{r_A X^2 + X(1 - X)}{r_A X^2 + 2X(1 - X) + r_B(1 - X)^2} \tag{12.37}$$

Note that all given equations are derived for bipolymers. (For a corresponding treatment of terpolymers from monomers A, B, and C, see, e.g. *Blackley DC, Melville FRS, Valentine L (1955) Proc Roy Soc A 227: 10*).

Molecular Weight (also: **relative molar mass**)
Dimension-less factor indicating the mass of a molecule as a multiple of 1/12 of the mass of an atom of nuclide ^{12}C. The numerical value of MW is equal to that of **molar mass** which indicates the mass in g/mol. The use of dalton as a unit of mass, identical with the atomic mass unit, is not recommended by the International Union of Pure and Applied Chemistry (IUPAC), see *Europ Polym J* (1984) *20 p. i–ii.*

Monomer Reactivity Ratios
Kinetic data defined as $r_A = k_{AA}/k_{AB}$ and $r_B = k_{BB}/k_{BA}$ where k_{AA} or k_{AB} are the velocity constants for the addition of a monomer A or B, respectively, to an A unit at the growing end of a copolymer molecule. Correspondingly, k_{BB} or k_{BA} refer to the addition of a B or A monomer to a B unit at the growing end.

Most Probable Chain-Length Distribution

$$H(P) = \frac{P}{P_n^2} \exp\left(-\frac{P}{P_n}\right) \tag{12.38}$$

where P is the chain length or *degree of polymerization* and P_n the *number average* of this quantity. Equation (12.38) is the limiting case of the *Schulz–Zimm distribution* for $a = 1$ and $P_w/P_n = 2$.

Net Retention Time, t''
Difference between elution time t_e and *dead time*, $t'' = t_e - t'$. In interactive chromatography, t'' can virtually become infinite, whereas in SEC the maximum value of t'' is equal to the elution time required for a column *pore volume.*

Normal-Phase Chromatography (antonym: *reversed-phase chromatography)*
Separation on a polar stationary phase, characterized by the increase of retention with the polarity of samples. NP gradients are gradients of increasing eluent polarity.

Number Average (see also: *weight average*)
The number average is the sum of products of relative frequency times individual value of the members of a distribution. For a MWD with n_i individual molecules of molecular weight M_i, the **number average molecular weight** is given by

$$M_n = \frac{\sum n_i M_i}{\sum n_i} = \frac{\sum m_i}{\sum (m_i/M_i)} \tag{12.39}$$

where m_i is the mass of constituents having individual M_i values. M_n can be calculated, e.g. from SEC results or measured directly by osmometry.

Pore Volume
Total volume of the pores of a column packing. In HPLC, the pores are usually filled with mobile phase. The total amount of mobile phase in a column is the sum of pore and *interstitial volume* (see Fig. 3.3). The ratio of pore size to solute diameter determines whether or not the molecules of a sample have access to the pores. Thus, in SEC the pore volume is the maximum range of elution volume available for separation of molecules of different size.

Random Copolymer (see also: *statistical copolymer*)
Copolymer with a distribution of monomer sequences which cannot be estimated from *copolymerization propagation probabilities*. Random copolymers can, e.g. be exotic varieties of *block copolymers* (with a comparatively high and unpredictable number of blocks). Polymers prepared from precursor homopolymers by incomplete chemical modification ("quasi-copolymers") may have also randomly arranged monomeric units but are not copolymers by definition.
Jenkins AD, Loening KL (1989) In: *Sir Geoffrey Allen (ed) Comprehensive polymer science, vol. 1, Booth C, Price C (eds) Pergamon, Oxford, p. 34*

Recursive Mayo–Lewis Equation
This equation enables the composition of a monomer batch to be calculated from the composition of a copolymer. The recursive form of Eq. (12.36) reads

$$G = \frac{(g-1) \pm [(g-1)^2 + 4g r_A r_B]^{0.5}}{2 r_A} \tag{12.40}$$

and the recursive form of Eq. (12.37) is:

$$X = \pm\left[q^2 - \left(\frac{x r_B q}{x - x r_B - 0.5}\right)^{0.5}\right] - q \tag{12.41}$$

where:

$$q = \frac{x(1 - r_B) - 0.5}{x(r_A + r_B - 2) + 1 - r_A} \tag{12.42}$$

Reversed-Phase Chromatography (antonym: *normal-phase chromatography*)
Separation on a nonpolar stationary phase, characterized by a decrease in retention with increasing polarity of a sample. RP gradients are gradients of decreasing eluent polarity.

Schulz–Zimm Distribution

$$H(P) = \frac{1}{c\Gamma(a+1)}\left(\frac{P}{c}\right)^{a} \exp(-P/c) \tag{12.43}$$

where $a = [(P_w/P_n) - 1]^{-1}$, $c = P_w - P_n$ and $\Gamma(a+1)$ is the *gamma function.*

Solute Acceleration Factor, *S*
Measure of the sensitivity of a solute to increasing elution power. The isocratic *capacity factor* decreases exponentially with increasing volume fraction of the stronger eluent, Φ_B, and can be approximated by

$$\log k' = \log k'_A - S \cdot \Phi_B \tag{12.44}$$

Here, k'_A is the value of k' in the poor eluent, i.e. at $\Phi_B = 0$. Equation (12.44) is useful for the description of *gradient elution* as well (adequate substitution of k' provided).

Statistical Copolymer (see also: *random copolymer*)
Polymer produced from more than one monomer species whose arrangement in the macromolecule is governed by the *copolymerization propagation probabilities* of the polymerization reaction. The common polymerization of monomer mixtures yield generally statistical copolymers.

Terpolymer (also: **Ternary copolymer**)
Copolymer produced from three species of monomers.

θ-Systems
Polymer solutions in poor solvents which cause second (and higher) virial coefficients to vanish. A behaviour of this kind requires a given polymer of infinite molecular weight in a peculiar solvent (θ-solvent) at a certain temperature (θ-temperature), θ-solutions behave like ideal ones.

van Deemter Equation
Relationship between the *height of a theoretical plate, h,* and characteristic data of a chromatographic set-up, expressing h as the sum of contributions from axial diffusion (varying with the inverse of velocity u), from the resistance to mass transfer (varying linearly with u), and from flow effects (e.g. eddy diffusion and substance delay in a mobile phase) with a more complex dependence on u. For comparison of different devices it is advantageous to introduce a reduced plate height $h^* = h/d_P$ and a reduced velocity $v = ud_P/D'$, which give

$$h^* = \frac{2\gamma}{v} + Av^{1/3} + C_i v \tag{12.45}$$

This expression, also referred to as **Knox equation**, indicates the contribution of axial diffusion, flow effect, and mass transfer as well as the importance of small particle size d_P and high values of diffusion coefficient D'.

Weight Average (see also: *number average*)
The weight average is the sum of products of weight fraction times individual value of the members of a distribution. For a MWD containing mass portions m_i of molecules with individual molecular weight M_i, the **weight average molecular weight** is given by

$$M_w = \frac{\sum m_i M_i}{\sum m_i} = \frac{\sum n_i M_i^2}{\sum n_i M_i} \tag{12.46}$$

M_w can be measured by, e.g. light scattering or calculated from SEC results through Eq. (12.46).

Author Index

Each entry indicates the reference number in the respective Chapter, e.g., 3-62 means Reference No. 62 of Chapter 3

Subject Index

(*Italic numbers* indicate Figures)